Sonic Fiction

THE STUDY OF SOUND

Editor: Michael Bull

Each book in The Study of Sound offers a concise look at a single concept within the field of sound studies. With an emphasis on the interdisciplinary nature of the topics at hand, the series explores a range of core issues, debates and objects within sound studies from a variety of perspectives and within a multitude of contexts.

Editorial Board:

Published Titles:

The Sound of Nonsense by Richard Elliott

Humming by Suk-Jun Kim

Lipsynching by Merrie Snell

Forthcoming Titles:
Sirens by Michael Bull
Sonic Intimacy by Malcolm James
Wild Sound by Michael Pigott

Sonic Fiction

Holger Schulze

BLOOMSBURY ACADEMIC
NEW YORK • LONDON • OXFORD • NEW DELHI • SYDNEY

BLOOMSBURY ACADEMIC
Bloomsbury Publishing Inc.
1385 Broadway, New York, NY 10018, USA
50 Bedford Square, London, WC1B 3DP, UK
29 Earlsfort Terrace, Dublin 2, Ireland

BLOOMSBURY, BLOOMSBURY ACADEMIC and the Diana logo
are trademarks of Bloomsbury Publishing Plc

First published in the United States of America 2020
Reprinted in 2022

Cover design and image by Liron Gilenberg,
www.ironicitalics.com

Bloomsbury Publishing Inc. does not have any control over, or
responsibility for, any third-party websites referred to or in this book.
All internet addresses given in this book were correct at the time of
going to press. The author and publisher regret any inconvenience
caused if addresses have changed or sites have ceased to exist,
but can accept no responsibility for any such changes.

Library of Congress Cataloging-in-Publication Data
Names: Schulze, Holger, author.
Title: Sonic fiction / Holger Schulze.
Description: [1.] | New York : Bloomsbury Academic, 2020. | Series: The study of sound |
Includes bibliographical references and index. |
Summary: "The first academic overview of one of the most advanced and controversial
approaches to sound studies, offering insight into its background, history, the present
discourse surrounding it, and its likely future impact"– Provided by publisher.
Identifiers: LCCN 2019028685 (print) | LCCN 2019028686 (ebook) | ISBN 9781501334795
(paperback) | ISBN 9781501334788 (hardback) | ISBN 9781501334801 (epub) | ISBN
9781501334818 (pdf)
Subjects: LCSH: Music–Philosophy and aesthetics. | Sound (Philosophy)
Classification: LCC ML3800 .S269 2020 (print) | LCC ML3800 (ebook) |
DDC 781.2/3–dc23
LC record available at https://lccn.loc.gov/2019028685
LC ebook record available at https://lccn.loc.gov/2019028686

ISBN: HB: 978-1-5013-3478-8
         PB: 978-1-5013-3479-5
       ePDF: 978-1-5013-3481-8
      eBook: 978-1-5013-3480-1

Series: The Study of Sound

Typeset by Integra Software Services Pvt. Ltd.
Printed and bound in the United Kingdom

To find out more about our authors and books visit www.bloomsbury.com
and sign up for our newsletters.

CONTENTS

Acknowledgments ix

Extradition: What Is Sonic Fiction? 1

1 Sonic Thinking:
   A Mixillogic MythScience of Mutantextures 19

2 Social Progress: Sensibilities of the Implex 43

3 Black Aurality: Alien Sonic Nontologies 61

4 Sensory Epistemologies:
   Syrrhesis and Sensibility 83

5 Acid Communism:
   A Haunted Utopia of Sound 105

6 NON: Ultrablack Resistance 123

Inconclusion:
Six Heuristics for Critique and Activism 141

Notes 152
References 158
Index 171

Acknowledgments

The idea for this book was born in several weeks in the spring of 1999, when I read the German translation of *More Brilliant than the Sun*. Since then I discussed, applied, transformed, and worked on the concept of sonic fiction in numerous talks, paper presentations, academic articles, book chapters, and course modules. The ideas then developed into the chapters that are now collected in this book. Along the way I received massive support through all the conversations with colleagues, young researchers and experts in this field, in focused interviews, workshop sessions, after paper presentations, in mail and chat conversations, and in other forms of academic exchange. Therefore my thoughts, my research, my arguments, and my form of presentation here are to an immeasurable degree indebted to the work, the thoughts, and the insights of Dietmar Dath, Detlef Diederichsen, Diedrich Diederichsen, Tobias Linnemann Ewé, Annie Goh, Steve Goodman, Rolf Großmann, Toby Heys, Macon Holt, Elena Ikoniadou, Johannes Ismaiel-Wendt, Sascha Kösch, Carla J. Maier, Thomas Meinecke, Pedro Oliviera, Malte Pelleter, Erik Steinskog, Jennifer Lynn Stoever, Achim Szepanski, and Christoph Wulf.

My special thanks goes to Caroline Bassett for discussing with me the best and the most appropriate translations for the various German language quotes in this book.

Last but not least a big thank you goes to Michael Bull for granting me the opportunity of publishing this title in this book series – and for ongoing conversation, support, encouragement, co-conspiration, and collaboration since our first encounter in 2006.

Extradition
What Is Sonic Fiction?

There is no sonic fiction.

There has always been sonic fiction.

Sonic fiction consists not just of one written account of sonic experiences and imaginations alone. Any small note, any aphorism or fragment of sound can qualify as a sonic fiction. Any small musical piece or skit on an album, any ever so small performative gesture by an instrumentalist carries at least certain remnants, trace elements, nuclei or mycelia of a very specific if not highly idiosyncratic sonic fiction. So too can any bit of liner notes or cover design, any bit of stage clothing, new instruments or pieces of software contribute to a sonic fiction. And, obviously, any gossip about performers or musicians, programmers or composers, fan extravaganza and upcoming new styles contribute to the ongoing and plastic, the malleable entity that one might call indeed: *a sonic fiction*.

Sonic fiction *is* everywhere. Where one can find sounds one will also detect bits of fiction. As a consequence sonic fiction might then *mainly* be found in the tiny and ephemeral, often rapidly vanishing intersections and interferences *between* texts and lifestyles, between a given recording medium, its material properties, its design and processes of storing, retrieving and reproducing sound – as well as all its listeners appropriating all these qualities of the recording medium to play an intrinsic and radiating part in their lives. A sonic, ephemeral fiction emerges between existing apparatuses for sound reproduction on the one hand and on the other hand all the idiosyncratic and incessantly transforming practices related to one material sound performance – be it live or recorded. A sonic fiction is just there. As soon as you listen, experience, digest or anticipate a given sound event, there are some germs of a sonic fiction planted in your sensory imagination, your reflection and desires. Sonic fiction is sensory sensibility.

> Now the story goes that Robert Johnson sold his soul to the devil at the crossroads in the DeepSouth. He sold his soul, and in return, he was given a secret of a black technology, a black secret technology, which we know now to be the blues. (*The Last Angel of History* 1996: 0:52–1:10)

Sonic fiction is, therefore, not at all ephemeral. It is not at all *merely* imaginary – even if the word *fiction* might attract such a notion, reduced to merely amateurish if not privatistic reflections; and even if the word *sonic* still might evoke some vague ideas of cryptic niche practices: 'At what point had the novel become such a small thing that it dwelt on the domestic problems of fictionalised characters?' (Kraus 2017). At what point had sound become such a small thing that it seemed only to be capable to represent exclusively privatistic urges and desires?

Sonic fiction is material and it is historical. Sonic fiction represents a thick cultural amalgamation of meanings and practices, sensibilities and techniques, represented not only in Kodwo Eshun's original *More Brilliant than the Sun* (1998) but also in John Akomfrah's famous visual essay *The Last Angel of History* (1996; cf. Gunkel, Hameed & O'Sullivan 2017: 249–267) or in Alexander Weheliye's concept of *Sonic Afro-Modernity* (2005). Two decades ago, in 1998, I encountered Eshun's book as a truly alien and generative artefact, bolted into the then contemporary discourse of the late 1990s; a time when I had just finished writing my first book on the modern history of aleatorics – and was just starting to conceptualize my second book on heuristics in the arts, in design, in music and sound (Schulze 2000, 2005). The intriguingly inventive German translation of Eshun's book by Marxist heavy metal expert and science fiction novelist Dietmar Dath – at that time already a longtime author for the conservative German newspaper *Frankfurter Allgemeine Zeitung* – was then its first interpretation that I got hold of. This alien scripture with a title and an author name both as alien as can be to me then, was introduced and praised in 1998 in *De:Bug*, a monthly journal and surely the most impactful, club-orientated and credibly intelligent publication on electronic dance music, remix culture, techno, and – in general – *'Digitale Lebensaspekte'*, all *digital aspects of life* at that time. What I read there *about* Eshun's book and later *in* Eshun's book intrigued me on an almost innumerable number of various multi-sited levels. Immediately, I got a hunch that this approach of *sonic fiction* could possibly open up a heuristic for artistic and for sonic practices that seemed promising and plausible to me. With sonic fiction as a foundation – so I imagined at that time – one would neither indulge in fake essentialisms and traditional truisms for analysing sounds or musics according to the latest positivist fad in the social or the natural sciences – nor would one be forced to a weird and truly unhealthy diet of fake analyses consisting of arbitrary Rorschach tests on occasion of listening to one's favourite record; the latter I knew all too well from the self-indulgent writings and ramblings of bourgeois representatives of an outdated, rancid and patriarchal thinking: a performance in words that apparently and sadistically loved foremost to abuse the arts as their lovely little pastime jester. Frankly, such conversations and rhapsodies about sound or music seemed to me just another tool to mark distinction and to fortify classist rejections and closures by means of being a connoisseur of decent descent, preferably from one of the ruling dynasties in economy, politics or academia. With Eshun's approach, though, writing about music and speaking about sound (Schulze 2006) could be – so I imagined – in general as convincing, as transparent and as consistent as possible because it remained stubbornly and self-confidently an inventive, individual and idiosyncratic heuristic.

When recently rereading *More Brilliant than the Sun* I was then surprised to note, however, that the term *afrofuturism*, which constitutes one main reference for the concept of sonic fiction, occurred only once in the whole book: in the first appendix to the book, an interview with its author, starting on page 175. In 1999 I might have simply ignored reading these appendices, but also on the previous pages the author preferred to play mainly rebelliously – so it seemed to me at the time – with the prefix *afro-*. All the struggles and the fights, the oppressive violence and the intrinsic motivations and contradictions of an afrodiasporic culture, its music, sounds and politics were represented in this book by precisely these rebellious and disruptive rejections of previous historical narrations and power structures; but never with an explicit reference or a scholarly introduction that a student like me then, in his late twenties, probably would have expected. To the contrary, Eshun coined the term sonic fiction as a new heuristic on the go, by means of heuristically proceeding through a large number of sounds and performers, of sonic experiences, imaginations and fictions that he writes about. Sonic fiction is introduced by way of sonic fiction. The imagination of a new heuristic for sound that was then triggered by sonic fiction in a reader who is of French-German-Scottish origin, now researching and living in Denmark and in Germany, stands to say the least in a strong contrast to these afrofuturist threads structuring, motivating and shaping what sonic fiction is. One of my main efforts in this book is therefore to explore sonic fiction as a black cultural concept with an intrinsically hybrid, politicized and revolutionary agency in an environment of still largely white endeavours in sound research: a cultural concept for the turmoils at the present time and all the transformations in the near future on this planet and beyond.

Sonic fiction was conceived by Kodwo Eshun as a concept on occasion of mostly afrofuturist cultural artefacts, of performances, musical compositions and sound pieces. However, it is one of the main characteristics of sonic fiction that it takes over traditions, practices and interpretations mostly outside of afrofuturism. This prolific, viral, contagious and assimilative quality of sonic fiction makes it a continuously inventive and transformative force, capable of generating differently crafted and new sonic, kinaesthetic and sensory fictions. With sonic fiction, I would claim today, afrofuturist knowledge and practices, arguments and historical re-narrations began to take over in the twenty-first century the existing hegemonic fictions of music and sound – in musicology and in music critique, in white musicology and white music critique, that is, with its 'overwhelming whiteness of scholars in the field' (Stadler 2015). For me, personally, sonic fiction provided and still provides guidance and provocation, a constant motivation in thinking and in writing not against but aside, underneath and beyond the truisms of common nonsense and sclerotic traditions. Meaningful explorations and their explosive research findings begin for me exactly with employing this very heuristic.

A Force of Liberation

Eshun's original book, in which he employs the term sonic fiction, started the ongoing conversation around this concept, inspired artistic, essayistic and academic appropriations of this term, this very book, *More Brilliant than the Sun*, never really defines its core term at one point. Sonic fiction is not proposed or even argued for as an instructive concept to tell artists, musicians or writers what they actually do. Or in the words of Eshun's famous claim in his introduction:

> In CultStud, TechnoTheory and CyberCulture, those painfully archaic regimes, theory always comes to Music's rescue. The organization of sound is interpreted historically, politically, socially. Like a headmaster, theory teaches today's music a thing or 2 about life. It subdues music's ambition, reins it in, restores it to its proper place, reconciles it to its naturally belated fate. (Eshun 1998: -004)

Eshun proposes, however, a reordering of the whole discourse. His goal in reordering is to avoid the superposition of self-indulgent, power-drunk and, lest we forget, still mainly *white theories* over the actually experienced bodily and technological practices to perform (to) this music. The discourse he starts then is not at all didactically explicating music or even covering it up with interpretations so familiar to protagonists of a largely white discourse; an addendum discourse that apparently can ignore quite easily the actual existence, practices and sometimes even prominent articulations by just those musicians and performers who play this music. Instead, Eshun approaches music and sounds by the means of energetically, mythically, and corporeally exploring them by touch, contact, interpenetration and amalgamation. The intrinsic polysensory and polyhistoric knowledge of music is fundamental to him, so:

> TechnoTheory, CultStuds *et al* lose their flabby bulk, their lazy, pompous, lard-arsed, top-down dominance, becoming but a single component in a thought synthesizer which moves along several planes at once, which tracks Machine Music's lines of force.

Far from needing theory's help, music today is already *more* conceptual than at any point this century, pregnant with thoughtprobes waiting to be activated, switched on, misused. (Eshun 1998: -004–003)

Eshun makes no effort to pedantically define his concept, then test its limits, fortify its borders and install a control system to administer what gets to be part of it and what needs to stay out. What he does instead is simply jump into writing by applying his new, still undefined and open coinage as a new framing, a new protagonist in the writing about music, a new *Denkfigur* (figure of thought). This new thinking object can then be filled and shaped and specified – and therefore also *factually defined* – in the course of usage as a new *single component in a thought synthesizer which moves along several planes at once*. This further development and definition by usage happens obviously to any new concept – but Eshun starts this process intentionally. He probes this term, tries it out, applies it, truly *essayistic* in the very sense of the word:

Stolen Legacy triggers the Egyptillogical Sonic Fiction of Earth Wind and Fire. Flip to the back cover of Shuzei Nagaoka's artwork for '79's *I Am* and there's the Egyptillogical landscape lit in the glaucous redlight of Dali-ized nuclear mysticism. (Eshun 1998: 156)

The new term *sonic fiction* appears in definitions that are more an inductive kind:

Both the name – 'Grandmaster Flash' – and the '81 track title – *The Amazing Adventures of Grandmaster Flash on the Wheels of Steel* – are Sonic Fiction. (Eshun 1998: 14)

In sentences such as these the ferment of sonic fiction operates as a *force of liberation*: liberating the writing, the thinking and the sensing of (not just about) music from scholarly restraints often superimposed on sonic experiences and imaginations of musical performances and productions. Writing sonic fictions – or even *PhonoFictions* – following the example of Eshun, means then unfolding those fictions inherent in cultural artefacts, musical productions and sonic performances:

Sonic Fiction is the packaging which works by sensation transference from outside to inside. The front sleeve, the back sleeve, the gatefold, the inside of the gatefold, the record sleeve itself, the label, the cd cover, Sleevenotes, the cd itself; all these are surfaces for concepts, texture-platforms for PhonoFictions. (Eshun 1998: 121)

It starts with the objects and inscriptions in which music is materialized. While writing and scrutinizing and narrating the sensorial, corporeal and personal effects a sonic fiction has, one can then move even further, into neighbouring realms, into connected narrations and meanings, semantics and imagery rooted in sensory experience:

Tracks become Sonic Fictions, sonar systems through which audioships travels at the speed of thought. (Eshun 1998: 25)

The *engine* of these *audioships* is *sensation transference:*

Sonic Fiction Is a Subjectivity Engine. (Eshun 1998: 121)

This *subjectivity engine,* this propelling force of sensation and imagination, accumulated, refined, distilled and stored in music opens up – no – it detonates in, it blows up the existing locations of musical experience and sonic imagination, in the midst of jailhouses, borderlines and fences:

Sonic Fiction replaces lyrics with possibility spaces, with a plan for getting out of jail free. Escapism is organized until it seizes the means of perception and multiplies the modes of sensory reality. (Eshun 1998: 103)

Sonic fiction is indeed a liberation force in the most precise sense of the word: a force to liberate epistemologies and historiographies, to liberate lifestyles and sensorial regimes, taste cultures and everyday practices – as well as styles of dancing and sounding, composing and performing, crying, squealing, howling and repeating:

Sonic Fiction is the manual for your own offworld break-out, reentry program, for entering Earth's orbit and touching down on the landing strip of your senses. (Eshun 1998: 103)

By sensation transference the audioship establishes a *Mothership Connection* – the main carrier and medium of John Akomfrah's film, referring to George Clinton's 1975 album with Parliament:

> Sonic Fiction turns your mind into a universe, an innerspace through which you the headphonaut are travelling. You become an alien astronaut at the flightdeck controls of Coltrane's Sunship, of Parliament's Mothership, of Lee Perry's Black Ark, of Sun Ra's fleet of 26 Arkestras, of Creation Rebel's Starship Africa, of The JBs' Monaurail. (Eshun 1998: 133)

Around the same time, when Eshun was working on *More Brilliant than the Sun* in the mid-1990s, Mark Dery introduced his famous series of interviews with authors and researchers Samuel R. Delany, Greg Tate and Tricia Rose with a troubling question:

> Why do so few African Americans write science fiction, a genre whose close encounters with the Other – the stranger in a strange land – would seem uniquely suited to the concerns of African American novelists? (Dery 1994: 179–180)

Encapsulated in this seemingly naïve and open-ended question is the core absurdity of alien lifestyles in an alien culture, i.e. black lifestyles in a seemingly non-black culture. This starting question for Dery's conversations then provided the ground to coin the new term *afrofuturism*:

> Speculative fiction that treats African American themes and addresses African American concerns in the context of twentieth-century technoculture – and, more generally, African American signification that appropriates images of technology and a prosthetically enhanced future – might, for want of a better term, be called 'Afrofuturism.' (Dery 1994: 180)

The liberation of sonic fiction is thus a direct, rebellious and dialectical consequence of life in such a deported, coerced, imprisoned and policed alien world of whiteness (Eshun 2005: 216f.). *We travel the spaceways.* Such a sort of *speculative fiction* is then provided with Eshun's *More Brilliant than the Sun*: a book that bears in its title a reference to one very specific, sonic amalgamation of then contemporary technoculture as drum machine scholar Malte Pelleter (2018: 35) recently pointed out:

Noticed that I was in this long dark tunnel, with a very, very bright light at the end, so brilliant … that was more brilliant than the sun. (Origin Unknown 1993)

It is at the very end of a track called 'Valley of the Shadows' that a female voice enunciating these very words does 'peel out of the slowly fading synthesizer arpeggio and finishes the sentence she had started again and again, only to be interrupted by sudden stumbling chops' (Pelleter 2018: 35;[1] translated by Holger Schulze) of a famous breakbeat from Lyn Collins's 1972 production of 'Think (About It)'. In a 'long dark tunnel, with a very, very bright light at the end, so brilliant … that was more brilliant than the sun' (Origin Unknown 1993) Eshun finds ways to:

Reverse traditional accounts of Black Music. Traditionally, they've been autobiographical or biographical, or they've been heavily social and heavily political. My aim is to suspend all of that, absolutely, and then, in the shock of these absences, you put in everything else, you put in this huge world opened up by a microperception of the actual material vinyl. (Eshun 1998: 179)

Instead of racialized biographies of musicians and social histories of music he opts for a more *multirational*, a sensorially materialist submerging into the whole *sensory spectrum* of *PhonoFictions* and all the *machine mythologies* actually in place here:

To say that today's producer is inarticulate and monosyllabic only reveals how standard criticism is deaf to the sensory spectrum captured in Sonic Fiction, PhonoFiction and machine mythologies. (Eshun 1998: 71)

Eshun captures this in:

A disconnected multirational Sonic Fiction, in which concepts jump, thought leapfrogs, mind zigzags from clause to clause, a perceptual current transmits between each intervals, ripples across gaps. (Eshun 1998: 43)

His goal is then, consequentially, not mere indulgence, pastime or an irresponsible or careless play with references, technological knowledge or sonic descriptions – like some reader indeed might have assumed, the goal of sonic fictions is – but precision, a precision though of higher complexities and meticulous sensibilities:

Yet in magnifying such hitherto ignored intersections of sound and science fiction – the nexus this project terms Sonic Fiction or PhonoFiction – *More Brilliant* paradoxically ends up with a portrait of music today far more accurate than any realistic account has managed. (Eshun 1998: -002)

Enforced Landianisms

The writing of Kodwo Eshun around the publishing year of *More Brilliant than the Sun*, 1998, took place in a constellation of writers, researchers, of sonic, of artistic and research practices connected to the somewhat pompously named Cybernetic Culture Research Unit (CCRU). The CCRU was allegedly founded in 1995 in the philosophy department of the University of Warwick, 'a dour, concrete campus set in the UK's grey and drizzling Midlands' (Mackay 2013). The group of people associated with the CCRU were initially gathered around theorist Sadie Plant – who left in 1997 to publish the cyberfeminist *Zeroes and Ones: Digital Women and the New Technoculture* (1998) – and Nick Land, who then took over the role as a sort of CCRU's patriarch, avatar as well as *spiritus rector*. While Nick Land is clearly a core author on the reading lists of the Alt-Right, neoreactionaries and neofascists in the 2010s, this further trajectory into an insanely antidemocratic and inhumane (not only post- or transhuman) eugenic hyper-racism was not clear to see in the late 1990s. Hence, this later development (and deterioration as I would argue) of him as a writer and thinker is not to be conflated with his earlier academic efforts at CCRU. Nevertheless, certain germs and nuclei of his fascist inclination might be found in his earliest explorations and ruminations. But, frankly, I do not wish to grant to the writings and activities of this white supremacist more space in this book on an afrofuturist core concept than seems absolutely necessary. Back to Warwick in the 1990s the members of the CCRU occupied, as Simon Reynolds recalled:

> An office on The Parade (Leamington's main street), rather than working c/o the Philosophy Department of Warwick University a few miles away... Inside CCRU's top-floor HQ above The Body Shop, I find three women and four men in their mid to late twenties, who all look reassuringly normal. The walls, though, are covered with peculiar diagrams and charts that hint at the breadth and bizareness of the unit's research. (Reynolds 2009)

Land fostered a widely transgressive approach at the CCRU. His thinking gravitated around rather extremist approaches of a darker and anti-humanist side of continental philosophy and art; most of those were only translated for the first time into English in the then recent years, the late 1980s and early 1990s – such as Antonin Artaud, Georges Bataille, Gilles Deleuze and Félix Guattari, Martin Heidegger, Jean-François Lyotard, Friedrich Nietzsche or Arthur Schopenhauer. In his writings and talks Land would then turn their borderline self-reflections and cocky world destructions, their quite daring demands into precise recipes for taking action, for introducing new academic formats, for intervening, breaking up and liquefying some of the congealed institutional rituals: 'theory was used as an element alongside music, art and performance' (Mackay 2013). For Simon Reynolds's later visit in the 2000s one performance at *Vibrotechnics*, organized by the CCRU in October 1997, was even re-enacted. In the tradition of tape looped *Lautpoesie* or sound poetry from the 1950s or 1960s Reynolds experienced this theory performance as follows:

> The first cassette-player issues a looped cycle of words that resembles an incantation or spell. From the second machine comes a text recited in a baleful deadpan by a female American voice – not a presentation but a sort of prose-poem, full of imagery of 'swarmachines' and 'strobing centipede flutters'. The third ghettoblaster emits what could either be Stockhausen-style electroacoustic composition or the pizzicato, mandible-clicking

music of the insect world. Later, I find out it's a human voice that's been synthetically processed, with all the vowels removed to leave just consonants and fricatives. Even without the back-projected video-imagery that usually accompanies CCRU audio, the piece is an impressively mesmeric example of what the unit are aiming for – an ultra-vivid amalgam of text, sound, and visuals designed to 'libidinise' that most juiceless of academic events, the lecture. (Reynolds 2009)

It might have been this *ultra-vivid amalgam of text, sound, and visuals designed to 'libidinise' that most juiceless of academic events, the lecture,* that provided a sort of meeting ground for both the writings of Kodwo Eshun and of Nick Land. Daringly erratic neologisms populate the writings of both authors; scarily dystopian and at times enthusiastically ecstatic narrations of future non-societies full of non-technologies for non-humanoids are the strange attractors for both of their essayistic thought experiments; both authors struggle for an expanded notion of theoretical writing that joyfully includes passages and paths into fictional narrations as well as intense imaginations of sensorial affects and disturbing sensations. As if over three decades later the poetic borderline visions of William S. Burroughs were rediscovered and repurposed as an academic method:

All music and talk and sound recorded by a battery of tape recorders recording and playing back moving on conveyor belts and tracks and cable cars spilling the talk and metal music fountains and speech…A writing machine that shifts one half one text and half the other through a page frame on conveyor belts…Shakespeare, Rimbaud, etc. permutating through page frames in constantly changing juxtaposition the machine spits out books and plays and poems – The spectators are invited to feed into the machine any pages of their own text in fifty-fifty juxtaposition with any author of their choice any pages of their choice and provided with the result in a few minutes. (Burroughs 1962: 64–65)

As a consequence, Eshun and Land are, probably most obviously, connected by their coinage of a new term for this new genre of expanded theory; a neologism that amalgamates two hitherto opposing genres of writing and allows for a massively accelerated, almost incessant genre-switching in writing and in thinking: *Theory-Fiction* (Land) and *Sonic Fiction* (Eshun).

Theory-fiction remains as undefined in Land's writings as is sonic fiction in Eshun's. Both authors prefer to guide their readers almost blindfolded into this newly proposed academic genre by means of its seductive qualities. The ingredient of *theory* is quite obvious in Land's texts – but where is the condiment of *fiction* to be found? The actual narrative passages in Land's *Thirst For Annihilation* (1992), *Dark Enlightenment* (2013) or in the collected writings called *Fanged Noumena* (2011) remain scarce if not non-existent and much less suggestive than the sonic fictions unfolded by Eshun. Because when Land is referring to fiction, he is less thinking of suggestively narrated novels and more of cleverly applied rhetoric strategies stimulating and tweaking the imagination of a theory's consumer: fabricating a fiction that feels so real and so present that it actually can have direct effects in real life as it provokes people to take action. Land and others trace these strategies back to Jean Baudrillard's influential essays on simulacra (Baudrillard [1981] 1994) and claim to apply them as a critique regarding contemporary culture and as a form of activism at the same time. Baudrillard's well-established theory from the high times of television culture and early media studies of the 1980s claims that in a media culture, most of the disseminated facts and documents are necessarily fabricated and refined to fit the daily transmissions and to enter everyday discourse: *ficta sunt facta*. Burroughs's 1960s poetic vision of a *Reality Studio* in which everyone's worldview gets fabricated was turned into a valid concept of the intellectual discourse: 'Storm the Reality Studio and retake the universe' (Burroughs 1961: 151). This very process of fabricating a fiction elaborating on selected documents, photographs, soundbites, names and persons can then be witnessed when reading one of the communiqués of the CCRU itself:

> CCRU retrochronically triggers itself from October 1995, where it uses Sadie Plant as a screen and Warwick University as a temporary habitat. ... CCRU feeds on graduate students + malfunctioning academic (Nick Land) + independent researchers + ... At degree-0 CCRU is the name of a door in the Warwick University Philosphy Department. Here it is now officially said that CCRU 'does not, has not, and will never exist.' (Communique

from the Cybernetic Culture Research Unit, November 1997, quoted in Fisher 2005)

Here, the loose association of some writers and researchers on a campus in the Midlands is indulgently exaggerated and reimagined as a transhuman entity. This institutional entity called CCRU then uses researchers *as a screen*, a university building *as a temporary habitat,* it *feeds on graduate students* and it even *does not, has not, and will never exist.* This self-mystification might be thoroughly tongue-in-cheek, but its rhetoric still conveys a lot more than just a factual description of an organizational unit. With such massively fictionalized statements a language operation is employed that previously had been analysed by Franz Koppe, German philosopher of language, in his treatise *Sprache und Bedürfnis* (*Language and Need*, 1977). This language operation will prove to be generative for both Land and Eshun (and many more) in their writings and thinking. In his book, Koppe adds one crucial and affective dimension to the common propositional analysis of texts. Usually, such an analysis would focus on the correct construction of a sentence and an argument in the philosophical sense – which can be true or false – with its subject, predicate, object and all the claims, the propositions it makes. To this so-called *apophantic substrate* ('apophantisches Substrat', Koppe 1977: 28–31), Koppe adds the affective dimension that constitutes obviously everyday life but often seems to be surprisingly missing in a propositional analysis of academic enunciations. This dimension he calls then the *endeetic surplus* ('endeetischer Überschuß', Koppe 1977: 97). This endeetic surplus frames the propositions made in a text with specific desires, needs, wishes and individual purposes, and also perspectivizations by its author. All texts that cannot be easily reduced to just a propositional calculus contain this mixture of claims and their framing, propositions and affects: a supposedly neutral if not bleak statement is therefore contextualized and complexified by its genuine situation of need – a *Bedürfnissituation* (Koppe 1977: 81). In the writings by Eshun or Land the propositions they hold true and wish to prove are not left on their own – but both authors add, in the words of Steve Goodman, a 'psychedelic function of theory' (Goodman, quoted in Fisher 2011).

It is this condiment of affects and needs that transforms therefore an otherwise blunt description of factoids on certain cultural phenomena into an erratically twisted, strangely shaped, narratively dense and vivid, energetic articulation of the political visions and cultural utopias of its authors. Theory becomes theory-fiction when its authors' *articulations of need* ('Bedürfnisartikulationen', Koppe 1977: 48–79) are mixed into, are represented by and stylized into these texts. These texts are not just sequences of propositions: 'It is ordinary language that's dumb and which must be adapted.' (Eshun 1998: 71) Their propositions carry a rich surplus of desires and dreams, imaginations and obsessions. Yet, both Eshun's and Land's articulations of need are substantially different from each other: *Afrofuturism* and *Neoreaction* might both react to cultural phenomena, but in an excessively opposite way. This very difference makes them incomparable if not existing in radically disconnected space-time continua of theory – including its enveloping social practices, historical processes and utopian urges. It is hard to conflate the struggle for a globally and interplanetary, intergalactically liberating and diversifying technoutopia of afrofuturism with the hope for a renaissance of a harshly aristocratic-oligarchic and decidedly supremacist dictatorship and a reinstating of an even more cruelly and violently ruling class in the visions of neoreaction. Even if both authors might have been present in the same university buidlings in Warwick in the same hours of a specific day at some remote point in the latest years of the twentieth-century, this truly arbitrary connection might have been the only one between them. Today, one might recognize the ambivalence and the scepticism when Eshun apparently asked at one point: 'Is Nick Land the most important British philosopher of the last 20 years?' (Fisher 2011).

More Like a Group of Otoliths

After the publication of *More Brilliant than the Sun* another author than Kodwo Eshun might have started to capitalize on this masterwork. He could have expanded this new approach to other genres, into other aspects of fiction and the sonic; he could have explored further subtleties of sensing, imagining and sounding in subsequent book publications, interviews, articles, exhibitions, in sound performances, maybe movies. Eshun did not follow precisely this path of application and capitalization. Instead, he shifted his activity to the wider array of the *ultra-vivid amalgam of text, sound and visuals* with the Otolith Group he co-founded with film-maker and social anthropologist Anjalika Sagar and 'an ever expanding group of artists, writers, thinkers and filmmakers to develop research, commission work and present ideas' (The Otolith Group 2018). To some of his readers this still seems like a turning away from academic publishing on music. However, one can also consider this an even more intensified entering of the ongoing discussion on sensory heuristics Eshun himself started with *More Brilliant than the Sun.* Eshun did not change tracks but the vehicle. This new one though has a life of its own:

> It coexists by curating, programming, publishing and supporting the presentation of artists work, contributing to a critical field of exploration between visual culture and exhibiting in contemporary art. (The Otolith Group 2018)

The Otolith Group, therefore, is a vessel that allows more profoundly and with more extensive journeys into the most remote trenches of sounds and the senses to continue these efforts *to 'libidinise' that most juiceless of academic events, the lecture.* With curated programmes such as *The Ghosts of Songs: A Retrospective of The Black Audio Film Collective 1982–1998* and *Harun Farocki. 22 Films: 1968–2009* (Tate Modern, London, 2009), video essays such as *People to be Resembling* (Haus der Kunst, Munich, 2012) or a solo exhibition such as *In The Year of The Quiet Sun* (Bergen, London and Utrecht 2014–15), this vessel proved capable of travelling some of the more remote water- and airways these days. But even the format of the academic lecture Eshun embraced and transformed recently as he held the first Mark Fisher Memorial Lecture in January 2018, precisely one year after Fisher had terminated his existence. In this lecture Eshun narrated and explicated not only his personal relation to the deceased thinker and inspiring university teacher – but he managed indeed to resignify the work of Fisher not only as that of an academic writer but of a supporter of new thoughts and ideas and even social groups of activism. In Eshun's understanding the people, the *humanoid aliens* (Schulze 2018), who come together and gather around explorative and imaginative specimens of writing and thinking, like the ones by Fisher (or Eshun himself, I might add) are rather peculiar ones:

> Those of us who are unable to reconcile ourselves to our existence. Those of us whose dissatisfaction and disaffection, whose discontent and whose anger and whose despair overwhelms them and exceeds them. And who find themselves seeking means and methods for nominating themselves, for electing themselves, to become parts of movements and scenes that exist somewhere between seminars and subcultures, study groups and HangOuts. Reading groups drawn together by the impulse to fashion a vocabulary. By a target. By a yearning. By an imperative to consent – in the words of Fred Moten quoting the words of Édouard Glissant – not to be a single being. (Eshun 2018a: 15:02–16:02)

Those of us that Eshun enumerates in a long list ranging from the *cybergoths* to the *sinofuturists* (Eshun 2018a: 16:44–22:38) are 'interpretive communities' (Eshun 2018a); or, as I would claim, they form *groups of otoliths*. What are Otoliths? The word otolith refers to the calcium carbonate crystals (or chalk) in the human but also other animals' inner ear (Highstein, Fay & Popper 2004). Your otoliths are *c.* 0.1 millimetres long, the ones of a sardine can be even *c.* 2.5 millimetres long (Dehghani et al. 2016). With these tiny stones, the *otoconia*, coupled to a set of hair cells in the inner ear, one performs a sense of orientation, of space- and time-based estimation of velocity and gravity in relation to listening, to one's body, to one's individual state of anatomy, physiology and corporeal sensibility. In your and my inner ears the so-called *utricle* represents and monitors the effects of largely horizontal motion, the complementary *saccule* then monitors the effects of vertical motion (Kniep et al. 2017). A group of otoliths therefore is physical matter that allows us to integrate dynamic movements sensibly in our actions and our lives. Precisely from your individual group of otoliths your orientation in space and time emerges: you are situated and you sense yourself as incorporated in the movements of these incredibly tiny stones. Your sensibility is, partially at least, embodied in these crystals. The Otolith Group by Eshun, Sagar and many others represent such an embodied sensibility, an *interpretive community.*

In this present book also the author, I, will write as such sort of an *otoconium*. As an embodied sensibility myself I will in the following six main chapters take you on an exploration of the effects and the transformations, the more legitimate or illegitimate appropriations, assimilations, domestications and reinterpretations of the initial concept of sonic fiction. What Eshun initiated with *More Brilliant than the Sun* indeed was a solar fusion that blinded and heated up and energized a large array of artists, writers, thinkers, musicians and researchers. The effects of this initial fusion can still be felt, sensed, registered across disciplines and cultures, various languages and across not seldomly conflictuous political and aesthetical approaches. Some of the authors and artists that I cite and scrutinize on the following pages might even reject my interpretation of their works being energized by Eshun's fusion – others, in contrast, might have more of a hard time being compressed and discussed together with some of the other writers, artists, researchers. This book is then not an introduction at all, that will close the case of sonic fiction once and for all in order to simplify its core concepts and issues to fit into a neat textbook. This is an *extradition*: it expels you maybe from your homeland and your homeworld – and it intends to propel you, way further away, into a *Black Atlantis* (Hameed 2016), into taking up the highly dynamic vectors of sonic fiction's energy, and to follow these trajectories. *Abducted by Audio. Possessed by Phono* (Eshun & Pomassl 1999: 5:33–5:39).

> What journey does the spirit make after it leaves the body? … Maybe this is how we can imagine the spirit's path: a solitary traveller exploring the stars, the only destination marked 'Further'. (Eshun 2005: 224)

1

Sonic Thinking
A Mixillogic MythScience of Mutantextures

This is a new continuum. You enter it right here and right now: a continuum of unheard terminology, of extraterrestrial locations and posthuman actors, connected through hitherto unknown threads, communicating and exchanging material carriers of information, of energy, and generativity in ways never dreamt of – with goals never thought of. A *mythscience* emerges here, as Eshun calls it in *More Brilliant than the Sun*, a new and apparently mythologically structured or grounded kind of scientific knowledge:

> I like reading books about John Coltrane, when he's sitting there studying music theory and he's listening to music from all over the world and trying to reach this higher order. I like the universe. (Eshun 1998: 92)

These words by Nathaniel Hall a.k.a. Af Next Man Flip are cited by Eshun because they articulate a genuine epistemological desire that lies quite transversal to any established form of commodified knowledge transfer, any university giving out certificates these days, any widely acknowledged research discourse. *Sonic thinking* starts right here: where knowledge is not mainly gained by academic reading, by discussing, falsifying or confirming, by rejecting or redefining propositions on some object called *sound*. Necessarily, any sonic thinking that merits this name has to start with sonic experience and by engaging in sonic writing, studying sonic sensibilities that are submerged in this experiential realm. Sonic thinking means to 'think with your ears' (Auinger & Odland 2007), to 'think *with* and *by means* of sound' (Herzogenrath 2017: 9). For sonic thinking the percept of:

> Sound is not merely yet another object for thought, taken in its limiting sense; rather, it is a demand posed to thought by that which it has yet been unable to think. (Lavender 2017: 246)

Sonic thinking therefore represents a truly 'paradoxical ambition *to think with, through and beyond sounds* all at once' (Schulze 2017: 218); it means to undertake rather 'a study through sound than a study about sound' (Papenburg & Schulze 2016: 1). So, does any sonic or *mythscientific* practice alone qualify already as a form of *sonic thinking*? Precisely this question was asked by its reviewers when *More Brilliant than the Sun* came out. Sascha Kösch, then editor of German monthly magazine *De:Bug*, for instance, concluded his review of the German translation of this book, *Heller Als Die Sonne*, with these words:

> Strangely, 'More Brilliant than the Sun' actually functions as a long review rather than as a theory book, or as a prime example of the application of various theories, which are also circulating in music itself. (Kösch 1999;[1] translated by Holger Schulze)

This *long review*, as Kösch put it twenty years ago, subsequently triggered a seemingly endless series of artefacts that are themselves again prime examples of the application of Eshun's theoretical framework. The concept of sonic fiction is now circulating in experimental sound pieces, in music theory, in sound art, in sound studies, and even in political theory. It has proven to be a concept not mainly for scholars or students of cultural research or musicology, but even more so for journalists and music critics – Eshun's main professions at the time of writing (e.g. Eshun 1992a, b, c, 1993a, b, 1995) – for cultural critics, for record lovers and club culture aficionados, for artists, inventors, and all sorts of thinkers and activists. This mix of all sorts of professionals and amateurs, of skilled and crafty persons who could possibly be affected and invigorated by Eshun's writing and thinking, this mix of activities points already to the second main concept contouring sonic thinking and sonic fiction alike – aside from mythscience: the concept of the mixadelic or the mixillogic.

This concept is close to earlier concepts of twentieth-century vernacular culture such as the *psychedelic, funkadelic* or *freakadelic*, also representing one possible diffraction from defined logics – be they 'Eurologics', 'Afrologics' (Lewis 1996) or *alienlogics* – such as *cartoon logic, magic logic, meta logic* like *transversal, three- or many-valued logic*, even close to *pataphysics*. Therefore a mixillogic is mixadelic insofar as it applies steps in thinking that would not be regarded adequate in scholarly logic following either antique syllogisms or contemporary rhetorics. Examples of diffracting logics are of major interest for contemporary artistic approaches of all sorts and they cannot be restricted solely to music, to installation art, concept art or theoretical reflections: they proceed mixadelically also in this respect – joyfully transgression being one characteristic trait of mixillogics. It is an ill and sick logic out of mixtures. When applying mixillogic, it might not be clear from the start what would be the outcome of this very mixadelic endeavour. A journey into sonic experiences and sonic thinking is foremost then a journey into:

MythSciences that burst the edge of improbability, incites a proliferating series of mixillogical mathemagics at once maddening and perplexing, alarming, alluring. (Eshun 1998: -004)

Maddening and perplexing, alarming, alluring quite precisely describe the common reactions to such transversal use and generative misuse of logics, bending and transforming them, adding to them, breaking and inverting them in unforeseeable ways. However, mythscience and mixillogics will at some point – according to Eshun – arrive at generating a certain kind of material traces, sonic traces maybe, that can be described with the third core concept of sonic fiction in *More Brilliant than the Sun*: the *mutantextures* made out of *mythsciences* and *mixillogics*. Mutantextures are not just mutated musical or sonic textures. Strictly following Eshun, a mutantexture results from the application of a divergent mixillogic. The emerging mutantexture – in sound or in any other artefact – is the actual and physical proof that a diffracted logic was applied. In reverse conclusion, it is very unlikely that one truly applied mixillogics if this practice only generated the same well-known and outworn textures, representing all traditional and reactionary cultural values and hierarchies. Mutantextures are the material proof of mixillogics in action and founded in mythsciences. In this first chapter on sonic fiction I will therefore explore the detailed effects and practices of these three major concepts of sonic fiction: *mythscience*, *mixillogic*, *mutantextures* being constituents of sonic thinking.

The MythScience of Sonic Warfare

A mythscientific history of all warfare known to operate by, through and with sound could jump back and forth between the year 1998, the year 2400, then into 403 BCE, stop by in 1677, 1738 or 1842, jump forward to 2020 and 2039 – and finally fall deep into prehistoric, even precosmic times of 13.7 billion BCE. Such a book on sonic warfare would exceed anything known in the traditional sphere of historiography: because it expands into timespans and time travels beyond any dimension accessible to humanoid aliens like you or like me and our eight to ten decades – if we're really lucky – on this troubled planetoid. Not one humanoid alien known to me or you could actually write such a mythscientific book about this unimaginably outstretched non-history; but still, one alien that goes by the name of Steve Goodman did write precisely this book. A book that – by standards of academic writing – exceeds many of the established and tacit, scholarly conventions and assumptions of many a reader. This book by the title of *Sonic Warfare: Sound, Affect, and the Ecology of Fear*, published by the MIT Press in the year 2010, sets in with a sensory fiction:

> *It's night. You're asleep, peacefully dreaming. Suddenly the ground begins to tremble. Slowly, the shaking escalates until you are thrown off balance, clinging desperately to any fixture to stay standing. The vibration moves up through your body, constricting your internal organs until it hits your chest and throat, making it impossible to breathe. At exactly the point of suffocation, the floor rips open beneath you, yawning into a gaping dark abyss. Screaming silently, you stumble and fall, skydiving into what looks like a bottomless pit. Then, without warning, your descent is curtailed by a hard surface. At the painful moment of impact,*

*as if in anticipation, you awaken. But there is no relief, because
at that precise split second, you experience an intense sound
that shocks you to your very core. You look around but see no
damage. Jumping out of bed, you run outside. Again you see no
damage. What happened? The only thing that is clear is that you
won't be able to get back to sleep because you are still resonating
with the encounter.* (Goodman 2010: xiii)

It is a nightmarish account, full of personal, all-too intimate
sensibilities – usually not appropriate to unfold in academic writing.
The sensible and poetic writing in this excerpt, unravelling a situated
entanglement, it enters rather laconically the assumed immaculate
clarity of an academic argument. Consequentially, this unsettling
narration of an intense sensory experience gets demarcated from
the rest of the text by a typographic marker: by italics. Surely, the
publisher, the author or the commissioning editor, they all hoped that
this typographic decision pushes this seductive narration a wee bit
more into the distance from the reader; after all, it should maybe not
contaminate the sacred realm of the argument too much. However,
Goodman develops his argument consistently through exactly such
a series of narrations and reflections, of such imagined and largely
fictional scenarios (grounded, though, most of the time in historical
research). The whole book is, substantially, a sequence of poetic
scenarios and epistemic imaginations that escort and provoke the
individual steps in the author's argument. Goodman outlines the
layered and intertwined structure of his book as follows:

The book is neither merely an evolutionary or historical analysis
of acoustic weaponry, nor primarily a critical- aesthetic statement
on the use of sonic warfare as a metaphor within contemporary
music culture.... Ultimately, *Sonic Warfare* is concerned with
the production, transmission,and mutation of affective tonality.
(Goodman 2010: xv)

His concern with *affective tonality*, though, is actualized in a
twofold way: in the object of his reflections – 'Sound, Affect, and
the Ecology of Fear' as it says in the subtitle – and in the way he
proceeds in his reflections, the μέθοδος or *method*: a series of fictions
affected by sonics; a sonic fiction. Goodman begins his first chapter
with a reference to John Akomfrah's video essay *The Last Angel*

of History (1996), a major audiovisual attractor for any reflection upon sonic fiction. In Goodman's study, though, sonic fiction is never discussed explicitly. However, he exemplifies, he executes, and he excels in it. Under the headline 'What is sonic warfare?' the author makes an effort to define his subject of writing:

> Finally, the sonic forms a portal into the invisible, resonant pressures that impress on emergent cyberspaces with all of their problematics, from virtuality to piracy. With increased online bandwidth, sound has attained a more central role in the polymedia environment of contemporary culture, unleashing unpredictable technoeconomic transformations resonating throughout global music culture. *Sonic Warfare* therefore also offers some insights into the economy of attention of contemporary capitalism. (Goodman 2010: 13)

The sonic forms a portal into the invisible, resonant pressures that impress on emergent cyberspaces with all of their problematics: this sentence leaves the ancient and often deserted edifices of academic writing and their strictly propositional language behind, chuckling cunningly. Goodman slams the door – and jumps onto Eshun's vessel. His bold, poetic and suggestive, imaginative claim (*'into the invisible, resonant pressures'*) on the effects (*'forms a portal'*) of a certain theoretical concept (*'the sonic'*) is not founded on definitions of this concept, the effects and this claim. A conventional scholastic argument would require this, at least. Instead, by leaving all definitions to the imagination of its readers, this writing style proceeds poetically, narratively, maybe aphoristically. It sketches, suggests, it expands on an already imagined scenario in the mind of its author – and then elaborates even more on the repercussions and consequences this imagined scenario might have (*'pressures that impress on emergent cyberspaces with all of their problematics'*). This writing is fictional and it is poetic. It is imaginative and suggestive, it is essayistic to a degree that its scarce non-essayistic portions become almost irrelevant. Goodman does not argue and then support his argument with empirical or historical examples, in order to finally interpret all of them to arrive at a desired conclusion. Goodman begins nevertheless with a statement in the form of an argument – but he jumps then right off as soon as possible into the narrative space of suggestive storytelling and poetic invention. He is narrating poetic concepts, he interweaves aphoristic reflections into the meshwork of selected propositional particles to in the end compose his *mythscience* of sonic warfare.

This writing style raises all the questions that recently have been and still are being discussed in the more advanced areas of humanities: how is it possible to integrate individual imagination and personal sensibility into research? What status of evidence could research of this kind then reasonably claim? What consequences would research of this sort have in the academic discourse? Doesn't it simply abolish all notions of objectivity, truth, of evidence or of insight? The analytical approach by Franz Koppe, introduced earlier in this book (cf. the previous chapter 'What is Sonic Fiction?'), though, lends us here a more precise set of terminology to understand what authors and researchers such as Steve Goodman are actually doing when writing this way. One might then ask: what need, what desire is articulated when academic language transgresses into fictional and poetic writing? It is, apparently, a materially affected writing that Goodman performs here – I'd even say: a *sonic writing* (Kapchan 2017, Schulze 2019b). Even more so, when he on the one hand explicates the current state of research in neighbouring research fields (e.g. spatialized sound reproduction, hyperdirected sound, weaponizing acoustic phenomena) – and on the other hand dives deeply into more speculative and imaginative forms of reflecting, arguing, connecting and even inventing future and past scenarios. Research in the technical and natural sciences on sound is his launch pad to project him to numerous concepts of cultural theory (e.g. by Friedrich Kittler, Henri Lefebvre, Jacques Attali or Paul Virilio), further on to artistic creations in literature, music and performance art (by J.G. Ballard, Public Enemy, William S. Burroughs, Underground Resistance) and lets him finally enter into even more daring areas of mythscience, he calls a 'black science' (Goodman 2010: 18). At this point, the concept of mythscience – aside from being an expansion of research conventions – obtains a truly generative if not explosive power (cf. Jasen 2016: 14): in Goodman's mythscience some highly idiosyncratic and almost pataphysical approaches from artistic and aesthetic theories cover the same ground and are discussed with the same earnestness as any other empirical research and then-recent developments in the engineering sciences. Research is expanded into imagination, into idiosyncratic sensibilities, and into predictive approaches.

At this point, considering the afrodiasporic origin of sonic fiction, it surely is no accident that Goodman speaks of this mythscience as *black science*. The whole discontinuum of political, social, artistic and revolutionary denotations and associations that such a naming carries, leads on the one side to a diffracting form of aurality: a *black aurality* (discussed in Chapter 3 of this book) – and on the other side it also implies a differing form of scholarship, including and embodying forms of resistance: an *ultrablack non-musicology* (discussed in Chapter 6). The alternate histories, boldly presented by Eshun or Goodman, are then, to say the least, also inspired by a desire to transcend the traditional linearity of historiography: the linearities and atomistic arguments of *white science* or *vanilla science*. These white historiographic narrations (White 1973) indulge primarily in eschatological developments of progress, superiority, ascent and bonhomie and are legitimated allegedly by continued dynasties of researchers and the royal houses of academic institutions. Eshun and Goodman, however, write their own diffracting, historically grounded but rather idiosyncratic 'MythSciences that burst the edge of improbability' (Eshun 1998: -004). Alternate mythsciences 'incite[s] a proliferating series of mixillogical mathemagics' (Eshun 1998: -004). One of these mathemagics is then Goodman's concept of *holosonic control*:

> Holosonic control operates through the nexus of directional ultrasound, sonic branding, viral marketing, and preemptive power…. It appears therefore that a major axis of sonic cultural warfare in the twenty-first century relates to the tension between the subbass materialism of music cultures and holosonic control, suggesting an invisible but escalating micropolitics of frequency that merits more attention and experimentation…. The micropolitics of frequency points toward the waves and particles that abduct consumers immersed in both the transensory and nonsensory soup of vibro-capitalism…. Because vibrational ecologies traverse the nature-culture continuum, a micropolitics of frequency is always confronted by strange, unpredictable resonances…. This vortical energetic terrain in the interzone between the artificial and natural environment constitutes the atmospheric front of sonic warfare. (Goodman 2010: 186–188)

This mythscience refers to an all-encompassing, sonically operating form of societal and governmental control; a cultural practice that indeed begins to unfold tangibly these days and to overtake the everyday life of many a *consumer citizen* (Schulze 2019a) all over this globe. Goodman, though, explored this development in his sonic fiction at a time when it was mainly imaginable in fiction, a decade ago. This careful procession into the impossible, the not yet known, the still seemingly irrational, this practice must then be regarded as a major motivation of any research – connecting the engineering sciences with a logic of Jazz, of the sonic or of afrodiasporic imaginations:

> MythScience is the field of knowledge invented by Sun Ra, and a term that this book uses as often as it can. A sample from Virilio defines it very simply: 'Science and technology develop the unknown, not knowledge. Science develops what is not rational.' (Eshun 1998: -004)

One materialization of this mythscience can then also be accessed in a record box set called *Martial Hauntology* (Goodman and Heys 2014): an experimental *audio paper* (Groth and Samson 2016) on sonic warfare, in which Goodman and his collaborator Toby Heys 'entwine imagined realities into conversations with history', foreshadowing even a video essay from the year '2056, when Corporations and Nation states have fused into single economic and political entities' (Ikoniadou 2016). The mixillogics of mythscience generated this mutantexture.

The Mixillogics of Sonic Epistemologies

Sonic thinking sets in with mythscience. How does it then proceed and explore sonic entities, expanding them into their embodied and sounding fiction? A mixture of approaches and sources, technologies and skills, practices and experiences come into play in the writings of Kodwo Eshun and Steve Goodman; they constitute the so-called *mixillogic* – a, well maybe, purposefully ill-advised logic of mixture:

> Breakbeat producer Sonz of a Loop da Loop Era's term skratchadelia, instrumental HipHop producer DJ Krush's idea

of turntabilization, virtualizer George Clinton's studio science of mixadelics, all these conceptechnics are used to excite theory to travel at the speed of thought, as sonic theorist Kool Keith suggested in 1987. (Eshun 1998: -004)

The 'dub virus' relates not just to the direct influence of the dub reggae sound on other musics but, more than this, its catalysis of an abstract sound machine revolving around the studio as instrument and the migration of a number of production and playback processes. The dub virus hacked the operating system of sonic reality and imploded it into a remixological field. The dub virus, taken in these terms, is a recipe for unravelling and recombining musical codes (Goodman 2010: 159).

Skratchadelia, turntabilization, mixadelics and the *dub virus,* they all represent *conceptechnics* that promote a specific knowledge, an embodied knowledge of practitioners and producers, of skilled and crafty persons, of artists. This knowledge is not *about* sound – it does not represent an auditory epistemology that could be extracted academically – but it is a knowledge out of, *'with, through and beyond sounds* all at once' (Schulze 2017: 218), a *sonic epistemology*:

Sonic epistemologies can be found in specific sociocultural fields in which practices dominate that are not (yet) established as relevant epistemic or even scientific practices. For the most part, these practices lack the reproducibility, the discreteness in documenting, and, therefore, the elegance that is topically postulated from relevant research practices. (Cobussen, Schulze & Meelberg 2013)

Sonic epistemologies *are* mixillogics. They constitute a body of knowledge that protrudes into a mixture of manifold, strangely formed and surprisingly combined practices. These mixillogic practices are not necessarily scholarly educated or executed according to the textbook – but they emerged out of a sonic sensibility in everyday practice and they generate, quite prolifically, sonic artefacts of many kinds. Sonic epistemologies, therefore, represent a form of *practitioners' theories, Praxistheorien, artists' theories, Künstlertheorien* (Lehnerer 1994). These theories are not written by scholars about artefacts generated by others – but they are the theoretical reflections and explications by these generators, these producers and practitioners themselves (Schulze 2005: 24–25), their 'Methododicy' (Lehnerer 1994: 8):

> Where and when is the decisive moment in every case when I no longer need to master the means (my knowledge and skills, my abilities), but bring them into play and let them go? ... How, according to which criteria do I then continue it? And when is it finished? (Lehnerer 1994: 147;[2] translated by Holger Schulze)

It is such mixillogics by practitioners in sound that form the core of *sonic thinking* (Herzogenrath 2017; Lavender 2017). Artists, practitioners and authors such as Salomé Voegelin, Brandon LaBelle, Sam Auinger or David Toop operate extensively in this new terrain. However, the academic status of mixillogical texts (and somewhat artists' theories) such as *Sonic Warfare, More Brilliant than the Sun*, also *Listening to Noise and Silence, Acoustic Territories, Hearing Perspective (Think with Your Ears)*, or *Ocean of Sound* is still not wholeheartedly welcomed by all scholars; especially not by those who favour and practise a more conventional approach to research, the *white science* as one might call it. Instead these *ill mixed artefacts* were easily assigned a so-called *special* and *extraordinary*, a *remarkable* or *distinct* place in academia. Such, at first glance, noble compliments, though were intended from the start to keep these irritatingly new, differing logics safely out of the main discourse of vanilla musicology or white cultural research. To praise them even more than what would be appropriate and polite should make clear: *this is definitely not an example of proper research. It might be interesting, inspiring, intriguing, maybe groundbreaking – but it surely is anything else than research.*

Regardless of this strong but concealed strategy of exclusion, more and more endeavours in mixadelic sonic writing were published; up to the point that the *Journal of Sonic Studies,* issued in Leiden, the Netherlands, decided in the early 2010s to dedicate a whole issue to these newly evolving mixillogics: 'Sonic Epistemologies' (Cobussen, Schulze & Meelberg 2013). The editorial to this special issue focused on two main problems sonic epistemologies face:

> How can we approach, analyze, and study sound? How can we disseminate our findings intersubjectively? (Cobussen, Schulze & Meelberg 2013)

At first sight a term such as *sonic epistemologies* might stand in harsh contrast to a concept like *mythscience*. Whereas the latter suggests a daring transgression into uncharted and illegitimate territory, the first term suggests more of an expansion of academically recognized knowledge: expanding the field of epistemology into the territory of sounds. One might then also assume that *sonic epistemologies* represent an academic overreach into non-academic areas whereas *mythscience* qualifies more as a non-academic overreach into an originally academic area. The first activity is expected, it is a well-known practice of, well, academic colonization and territorialization; the latter, however, is truly a breach of conduct, a subversive if not revolutionary act. Both of these heterogeneous movements, though, meet on the same ground as soon as their individual goals are effectively reached – that is: as soon as mythscience and sonic epistemologies both establish their mixillogics as a common area of knowledge hitherto unknown. The colonialist undertone of this territorialization, though, remains – and the critique of a *white aurality*, performing a strong desire of such territorialization and colonization, quite convincingly articulated recently by Annie Goh and Marie Thompson (Goh 2017; Thompson 2017), will be discussed in more detail in Chapter 3 of this book on *black aurality*.

The ambivalent impression regarding sonic epistemologies, however, might still remain when investigating the underlying theoretical framework: the approach of *sonic materialism*, so prolific and stimulating in sound studies recently. To this larger strand of research the writings of Goodman, Eshun, but also other writers mentioned earlier in this section certainly belong (and surely my own writing also can be considered part of this strand). Sonic materialism has been defined by two articles: one by Christoph Cox ('Beyond Representation and Signification: Toward a Sonic Materialism' in an issue of the *Journal of Visual Culture* from 2011) and one by Salomé Voegelin ('Ethics of Listening' in an issue of the *Journal of Sonic Studies* from 2012). Both articles represent and request specific efforts in research concerning sonic epistemologies,

> In favor of a new sonic ontology in which the current aesthetic theories concerned with representation and signification should be replaced by a sonic materialism, and a sonic realism should dispel an anthropocentric idealism and humanism. (Cobussen, Schulze & Meelberg 2013)

These efforts to transcend anthropocentric notions take thus a Deleuzian and Spinozian road into dynamized and unstable materialities, the plasticity and malleability of actors; their agile activities, concepts, habits and perceptions are recurrently underlined:

> Instead of fixed identities and meanings, stability, nouns, and stasis, the sonic exposes us to action and movement, to fleeting understandings, verbs, and contingent possibilities. The ear's focus is on process, on objects and events existing in time. A sonic materialism is a temporal materialism, grounded in a contingent encounter of listening. (Cobussen, Schulze & Meelberg 2013)

Salomé Voegelin goes even one step further:

> Sound's ephemeral invisibility obstructs critical engagement, while the apparent stability of the image invites criticism. Vision, by its very nature assumes a distance from the object, which it receives in its monumentality. Seeing always happens in a meta-position, away from the seen, however close. And this distance enables a detachment and objectivity that presents itself as truth. Seeing is believing… By contrast, hearing is full of doubt… Hearing does not offer a meta-position; there is no place where I am not simultaneous with the heard. However far its sources, the sound sits in my ear. I cannot hear it if I am not immersed in its auditory object, which is not its source but sound as sound itself. (Voegelin 2010: xi-xii)

Apparently, a very strong *sonocentrism*, even a taste of the old and convincingly deconstructed *audiovisual litany* (Sterne 2012: 9; Schrimshaw 2015: 159–160) can be detected right here. Where does this almost moralist tone of superiority and only slightly concealed uninhibited praise of one sensory approach, one bodily sensibility and one cultural practice come from? Is this just the well-known boasting and self-praise of an (frankly, not really any more) underdog, an outlaw, a freak? Rightfully, hence, Will Schrimshaw exposed the 'idealised hearing and apparently universal "sonic sensibility" constructed in accordance with a nature or metaphysics of sound in opposition to visuality' (Schrimshaw 2015: 159) dominating these texts. His critique is spot on in detecting the sonocentrism in both examples. However, his goal to excavate a coherent anti-rationalist, technophobic and non-textual front within sonic materialism leads him to identify falsely Cox's proposal of an 'anonymous sonic flux' (Cox 2011: 155–157), a 'sonic philosophy' (Cox 2013, 2018) or Voegelin's notion of a 'sonic sensibility' (Voegelin 2014: 13, 24) with an iron-clad sonic essentialism that implies transcendental and metaphysical truths derived from sounding and listening. How Cox and Voegelin unfold the repercussions, the malleable, and also the idiosyncratic sensibilities when experiencing sound, that defies actually such an idea of a consistent essentialism: their writings are performing anti-essentialism consistently. Though, and that might be the main point of attack, also hunches and imaginations of essentialism might even enter their reflections on the,

> *Sonic flux*, that is the notion of sound as an immemorial material flow to which human expressions contribute but that precedes and exceeds those expressions. (Cox 2018: 2)

> Sound's purposelessness is not its irrelevance or non-intentionality. Listening and sound making are highly intentional and generate their own contingent purpose. (Voegelin 2014: 114)

In one word: Schrimshaw seems to recognize here the missionary and self-stylized *audiopietists* (Schulze 2007, 2018: 230, 2019a: 202–208), that populated early sound theories by Raymond Murray Schafer over Joachim Ernst Behrendt and that still perform their act as truisms for sonic evangelists especially in sound branding or sound art.

Yet this, well, caricature of true sonic believers preaching their catechism does not really apply to the aforementioned authors, Voegelin, Cox or others, it does not apply to their main writings, and not to their research and teaching practices – though Schrimshaw wishes to apply it. The mixillogic and the diffracting sciences in their writing, performing and thinking are way too strong in them. Even, again, if they might also include trace elements of essentialism as perceptual effects and convincingly integral parts of a sonic experience, now and then: not to erase these trace elements qualifies in my reading as a form of convincing source critique and academic rigour in representing the full range of sensations in a sonic experience.

Forms of sonocentrism, also a praise of *epiphanic sonic experiences* can be detected in sonic epistemologies and also in mixillogics – though interwoven with heterogeneous other sensibilities and figures of thought. However, the underlying conflict and implied dissent to which Schrimshaw is reacting here seems more to be a conflict between artistic explorations and practitioners' theories of sensibilities, of hunches and senses, of yet unclear, malleable, evolving and transforming concepts on the one side (represented by Cox, Voegelin and others) and the academic and professional review and analytical critique of concepts, terminologies, skills and practices, categories and dispositives (represented by Schrimshaw) on the other. In a nutshell, this resentment and conflict is also a materialization of the different cultural practices (and subcultures) centred around propositional or discourse analysis as well as a highly competitive debate culture on the one side and the skilled practices of *syrrhesis* (Serres) or *mixillogic* (Eshun) as lived and experienced by cooks and DJs, musicians and tailors, painters and video rendering specialists on the other. Though a lot of protagonists are present in both lifestyles, both forms of habitus and both fields of profession, nevertheless, these two fields can be rather rigid in excluding distinct performances from the other field as inconceivable, as ridiculous, as simply irrelevant. To this very exclusion of practice, of sensibilities, of flesh and materiality, some more outspoken antagonists of sonic materialism and sonocentrism often react; and this exclusion often is then executed by basically denying an inherent material logic, a mixillogic that guides practitioners and artists and designers and skilled persons.

Consequently, if sonic epistemologies are to be taken seriously, it is necessary to ascribe to those alternate, thoroughly sonic forms of knowledge the same dignity as ascribed to forms of knowledge that are easily transferred into discrete and reproducible, semiotic and alphanumeric codes, easily functionalized and commodified in contemporary consumer culture as well as in industrialized research. Or, in Eshun's words:

> Music is heard as the pop analysis it already is. Producers are already pop theorists. (Eshun 1998: 004)

This liquefying of epistemologic discourse and this re-entry of artistic artefacts as actual epistemic articulations is – psychologically speaking – apparently unsettling to not a few scholars and thinkers: an effort to rehabilitate producers' mixillogics by stressing their epistemic impact apparently must provoke them to fervently rebuke those very producers. The mixillogic of a producer's theory is often hard to swallow with its offensive hypertrophy of articulations of need (Koppe), of their *endeetic substrate* over its propositional substance:

> The object as thing is an activity, it *is* to do: being as the production of possibilities rather than the appearance of totality. (Voegelin 2012)

This 'sonic flesh' (Voegelin 2014: 127) constitutes for sonic materialism and sonic epistemologies alike a,

> Contingent body of perception, the 'sensible sentient' that sees and hears not a positive, transcendentsal object separate from itself, but perceives things through their common simultaneity within the world. The fleshly body sees things through being seen and touches itself touching others. (Voegelin 2014: 128)

This indeed is a major provocation for disembodied academic research still claiming to apply,

> A presumably anonymous, generalizable, and ahistoric research practice with outcomes of a similar nature. A supposedly total abstraction of desires, obsessions, affects, and imaginations of individual researchers. (Schulze 2018: 12)

These assumptions run contrary to mixillogic and material, sonic epistemologies. The benefit of sonic epistemologies is to materialize indeed forms of mixillogic knowledge that are primarily, accessible via the auditory, to expand the universe of known epistemic practices into existing mythsciences, and to transcend and transform, therefore, also the logocentric epistemologies of the white sciences and white auralities. Mixillogics give room to the very specific sensory approaches of truly alternate experientialities, of alternate forms of existence of alternate subcultures and idiosyncratic biographies – with their very own particular sensibilities inscribed and embodied in their flesh (cf. Cobussen, Schulze & Meelberg 2013): embodied mutantextures.

The Mutantextures of Sonic Possible Worlds

Practising mixillogics on the ground of mythscience will generate diffracting artefacts also – differing kinds of sound pieces, different specimen of texts: *mutantextures* emerging out of mythscience and mixillogics. Eshun writes:

> Between '68 and '75, Macero & Miles, Hancock *et al* turned effects into instruments, dissolving the hierarchy by connecting both into a chameleonic circuit which generated new mutantextures. (1998: 42)

> Skratchadelia are mutantextures generated by turntabilization, by using the turntables as universal tone generators. (1998: 43)

Apparently, generating mutantextures is the most prominent goal when employing mixillogics. But what precisely is achieved when a mutantexture emerges? Eshun developed the concept of sonic fiction and its implied concepts of mythscience, mixillogics and mutantextures as means to open up contemporary discourses in cultural studies for the then still rejected and repressed mythsciences of afrofuturism. Authors, artists and researchers took up his original concept and repurposed it more and more – the writings by Steve Goodman and the use of sonic fiction in sonic epistemologies are both examples of this. This process of appropriating a new concept, of including a hopefully creative misreading, then resulting in a repurposing and a specific redefining of the original concept, all of this is rather common practice in research. Concepts, approaches and methods are not the property of one inventor, researcher or author. As soon as they are out in the public sphere of research and of thinking, of design or of artistic practice, they surely will be applied, misappropriated, reinvented, repurposed and used in alien contexts. Even thoroughly wrong and unsettling misappropriations need, from the perspective of critique, to be recognized as basically legitimate appropriations. However, a concept that is so rooted in a specific and politically as well as historically loaded discourse – in this case afrofuturism and black diaspora – poses a challenge if applied to new contexts and in altered ways. It is not just any ahistorical and context-free entity that could be applied and used in any possible way. The history of colonization, of territorialization and illegitimate misuse and misappropriation constitutes an inherent part of it – so precisely these practices, if apparently applied, need to be scrutinized with even more rigour than already well established. It needs respect, a radical imagination and a sort of openness towards the unexpected to apply sonic fiction in an appropriate way. Eshun himself made this at least a tiny bit easier, because he purposefully did promote this concept not to be enclosed in a gated community of discourse participants but to be opened up, to be applied and repurposed in a wider, maybe the widest discourse. More mutantextures ensued.

Salomé Voegelin, for instance, took this conceptual tool to open up contemporary discourses in cultural and sound studies for new mutantextures. These are generated here through the mythscience of *idiosyncratic sensibilities* – sonic sensibilities, corporeal sensibilities, illogical sensibilities. In her writings the concept of sonic fiction retains its major, generative function. She explicates her attachment to this concept in a footnote pinpointing the major difference to Eshun yet acknowledging their shared goal:

> The term 'sonic fiction' is reached via a different route and crossing different references, but it nevertheless shares in description and conviction with some of Kodwo Eshun's ideas as articulated in his book *More Brilliant than the Sun*. Like his sonic fiction mine too '…lingers lovingly inside a single remix, explores the psychoacoustic fictional spaces of interludes and intros, goes to extremes to extrude the illogic other studies flee. It happily deletes familiar names […] and historical precedence.' (Voegelin 2014: 183)

And further on she marks the difference of her approach by a decisively corporeal access to sonic fiction:

> My sonic fiction lingers in the illogical found via the body listening rather than in history and canonical names, to ignore 'comforting origins and social context' and build contingent ones instead. But it does so via literary evocations and as possible worlds rather than as science fiction. (Voegelin 2014: 183)

This footnote can be found in Voegelin's *Sonic Possible Worlds* from 2014. In this book she explicitly lays out the process by which sound allows for a mixillogic expansion of the mythscientific imagination of a listener into the mutantextures of highly idiosyncratic and sonic possible worlds:

Sound does not propose but generates the heard whose fictionality is thus not parallel but equivalent: it produces a possible actual fiction rather than a possible parallel fiction and sounds as 'world-creating predicate.' Sonic fictions do not propose a bridge between the actual and the possible but make the possibility of actuality apparent, building reality in the contingent and rickety shape of its own formless form. Thus, the sound artwork as sonic fiction is a phenomenological, a generative fiction, rather than a referential fiction. It is designed from the actions of its own materiality, not as description or reference of an object, a source, but as sound itself; we inhabit this materiality intersubjectively, reciprocating its agency in the sensory-motor action of listening as a movement toward what it is we hear. (Voegelin 2014: 51)

Sonic fiction is a phenomenological, a generative fiction, rather than a referential fiction: Voegelin starts out her exploration with the mixillogical in sensing sounds and sonic fictions – and moves then into, what she calls, a 'phenomenological possibilism' (Voegelin 2014: 48): sonic possible worlds being triggered by sensory experiences. She writes, therefore, an audile phenomenology of *mutantextures*:

Writing about the possibility of sound is a constant effort to access the fleeting and ephemeral, that which is barely there and yet the influences all there is. (Voegelin 2014: 2)

Voegelin stresses the highly dynamic, plastic, situated and relational character of sound events and the sonic experiences of you, me, of his or hers. This might again provoke suspicions of sonocentrism or audiovisual litany – yet, her approach connects more to the mythscientific strand of sonic fiction, generating an almost endlessly deviating plurality of possibly conflicting forms of sonic knowledge, of worlds and *life-worlds*:

> The universe I want to draw on is not centered around and constructed from one world only, but is constituted of a plurality of actual, possible, and impossible sonic worlds that we can all inhabit in listening and through whose plurality music loses its hegemony and discipline and the landscape gains its dimensions. (Voegelin 2014: 14)

Both authors, Voegelin and Eshun, promote a broadening of the spectrum of accepted forms of knowledge; both increase in their writings the contingencies in their approaches, the perspectives, epistemologies and ontologies. Eshun increases these in direction of formerly apocryph, electronica-born and deviating afrocentric aesthetics, Voegelin in direction of formerly considered idiosyncratic, sensibility-related and often repressed *hypercorporeal* aesthetics (Schulze 2008):

> The possible worlds of Descartes and Leibniz, considered through a sonic sensibility, are not determined by God or by science, which are not its necessity, the bearer of its reason and truth. Instead, sonic possible worlds are 'chosen,' as in generated, by the listener and reveal the contingent possibilities, sonic 'extensions,' of actuality in which they take part not through a 'negation if negation' but through negotiation between your invisible world and mine. (Voegelin 2014: 24)

A hitherto fixed and metaphysically ordered selection of propositions and episteme, methods and arguments – not seldomly in reference to belief systems promoted by white, male, Western, Christian doctrines and dogmas – are replaced therefore with a more mobile and malleable set of constituents. Voegelin and Eshun set them in motion, they dynamize, relativize and connect them, they corporealize, materialize and amalgamate them in new and unforeseen constellations. Hypercorporealism or afrofuturism appear as somewhat implicit goals of their mythsciences, and mixillogics guide them to the mutantextures both books, *More Brilliant than the Sun* and *Sonic Possible Worlds*, represent as written and printed, as material objects. Voegelin even explicitly rejects the choice of one finite ontology – for example an *Ontology of Vibrational Force* as proposed by Steve Goodman (2010: 81–84). One might find, though, in the following words of Voegelin a rather mixillogical ontology of a sensorial continuum of sound – a, if you will, *mixillontology*:

> The absence of an actual ontology, replaced by a plurality of non-hierarchical histories as anecdotes and contingent connections that do not reveal an a priori but generate their own secrets, and the fact that these possibilities exist in 'closeness', as possibilities of one sonic universe, makes a joint critical framework for music, sound art, and the acoustic environment possible. The paradise of a sonic possibilia allows us to hear a continuum of sound that neglects disciplinary boundaries to sound, music, the soundscape, and sound art as close worlds and gives us new insights into the possibility of the world of which they all are variants. (Voegelin 2014: 145)

This *continuum of sound* that can be approached with 'critical immersivity' (Voegelin 2014: 124) and a 'phenomenological impossibilism' (Voegelin 2014: 158) constitutes the mutantexture of sensibilities in Voegelin's understanding:

> In this sense a phenomenological impossibilism performs a primacy of perception that reveals the rationale … of that which is possibly not existing but is nevertheless imaginable, and of that which is not imaginable but nevertheless existing, the impossible, all of which play a part in the plural possibility of actuality. (Voegelin 2014: 158)

Goodman and Voegelin present complementary and not seldomly conflicting interpretations of and further elaborations of sonic fiction. Goodman's technoimaginative exploration of sound, affect and the ecology of fear across history and across the sciences connects here transversally and dialectically with 'a tuning into the world in order to see all it could be … through the plurality of a sonic sensibility' (Voegelin 2014: 13). Starting with Eshun's sonic fiction both authors indeed reverse and revolutionize the antique and sound theory as represented by Raymond Murray Schafer and some members of the World Soundscape Project. Voegelin,

Goodman and other protagonists of sonic materialism and sonic thinking propulse these sonic theories through mythsciences and mixillogics into the mutantextures of the twenty-first century. With sonic fiction, the transdisciplinary and progressive research through sound accelerates to match and to challenge the speed and the complexity of everyday sonic experiences in the present and, supposedly, in the near future. How will you conceptualize, narrate and analyse the mythsciences of the 2040s, the mixillogics of the 2070s? Or the mutantextures of the 2120s?

What Is Sonic Thinking?

Sonic thinking – according to Kodwo Eshun's approach of sonic fiction and some of his interpreters in action such as Salomé Voegelin and Steve Goodman – can be centred in the midst of three radiating nuclei. Some approaches will gravitate more to one of them, others will oscillate between two, some will stay static or move incessantly between all three. These three core concepts are mythscience, mixillogic and mutantextures. These resources of deviating knowledge, of epistemic practices, and of textures of artefacts provide the potential to engage in sonic thinking and, consequentially, to expand, to elaborate, or to unfold a sonic fiction. In what ways are these three forms of knowledge, practices and artefacts now deviating precisely?

Mythscience, mixillogic and mutantextures diffract the white sciences of knowledge, practices and artefacts, so they can move away from the more linear trajectories of logocentrism, of established political, social and historical hierarchizations and commodifications, as well as from guiding frameworks and *grands récits* such as the narrative of progress. As three generative nuclei they achieve this reordering of an established continuum of epistemology, of thinking and of research by a set of transformational questions. The resulting mutantextures as well as the proceeding by mixillogics and the resource in mythscience transform altogether the relevant epistemologies with the question: *How do we think beyond logocentrism?* (Schulze 2017: 228–233). The conventional logocentric argumentations and debate rituals as well as the obsessive and highly idiosyncratic focus on writings and the practices of writing cultures are being expanded into the wider area of generativity, including then all sorts of experiential and performative means of expression. With this expansion research and thinking enters differing material continua. Sonic thinkers might ask: *How do we think corporeally? How do we think spatially?* (Schulze 2017: 224–228, 220–224). Moving away from meticulously crafted textual character strings and into the realm of performativity, sensibilities and corporeality, the wider variety of idiosyncratic and tangible interferences and interpenetrations between the related sonic generators and protagonists turn into the structuring forces of thinking and epistemology. This entails an expansion into the intricate details of all the historically, culturally and materially determined, and thus highly situated and intrusive conditions of any sonic experience. Sonic thinking and sonic fiction are gleefully heteronomous approaches to sounding and imagining. Finally, all of these expansions of conventional forms of thinking and epistemologies lead to a transgression that might be the hardest to accept for academic writers: *How do we think imaginatively?* (Schulze 2017: 233–237). The format of sonic fiction leads its protagonists, writers and inventors to an imaginative thinking as a method to confer sonic experiences by means of a poetic or narrative immersion with more erratic, surprising and unconventional forms of performativity. This writing transcends then radically the focus on proposition and argument; not only does it integrate narrative passages but at times it favours erratic articulations of need and desire over the orderly disposition of reasoning efforts:

This writing is a soundscape composition (Voegelin 2014: 13).

It produces a sonic philosophy that scrambles the separation between theory and its musical objects of study. In this way, it still stands as one of the strongest examples of Eshun's suggestion that electronic music has no need to be rescued or theorized by a transcendent cultural theory but is instead already immanently conceptual. (Goodman 2010: 160)

All the ideas seemed to rush towards this – sonic fiction seemed to be an attractor – and all the terms just moved towards it and it was the easiest thing in the world to extract them and plug them all into each other. (Eshun quoted in Weelden 1999)

2

Social Progress
Sensibilities of the Implex

Kodwo Eshun's book was released in 1998 by Quartet Books in London – and it was soon out of print. Already in September 1997, at the famous *Loving the Alien*-conference, organized by Diedrich Diederichsen at the Volksbühne Berlin (Diederichsen 1998), Dietmar Dath met Kodwo Eshun for the first time, he heard of this upcoming book – whose German translator he should become soon after. The German ID-Verlag from Berlin and its head, Andreas Fanizadeh, then approached Dath to translate this volume into German. He translated it in roughly two months (Dath 2018b).

Dath is an important figure, a prolific writer and a widespread erudite intellectual in Germany since the early 1990s. Since 2007, with only a brief hiatus, he has written for the conservative newspaper *Frankfurter Allgemeine Zeitung*. He publishes novels, mostly outspoken science fiction or at least with a strong twist into the science fiction genre; he is an expert in heavy metal and Marxist theories alike, and together with the three members of the free float jazz ensemble Kammerflimmer Kollektief, he publishes songs and album records under the band name of The Schwarzenbach (2012, 2015). His translation of *More Brilliant than the Sun* was a stunning effort and actually a massive boost for the discussion of afrofuturism and black diaspora in the German-speaking world. But no one did surely foresee the long-lasting and truly wider impact of this translation: the German translation of *More Brilliant than the Sun* was for a longer period of time the only printed version one could buy – aside from all the scans and PDFs cleverly hidden and provided in the wilder archives of the global networks. This book and especially this translation inspired a larger number of younger German researchers, artists or dedicated aficionados of all sorts of sound art and sound productions to dig deeper into the issues and the trajectories, the struggles and the glorious artefacts of afrofuturism and all the related traditions discussed in this book.

Dath's translation of Eshun's book has had, therefore, an impact that is not unusual to observe in the publishing history of academic titles as well as in the history of fiction or even poetry. The study of comparative literature across the limits of one individual language and its community of readers documents time and again how only a valid translation of a crucial text can indeed provide its actual impact in the new language. Whereas the original text might more often submerge in the mass of published texts of the same kind, only recognized and read by the experts and the diehard fans, the translated work now and then factually makes a difference: for the wider community of readers these texts only appear on the surface of potentially interesting publications and cultural artefacts as soon as they are translated – every time anew a shocking event in their culture. In this case, the work of translator Dietmar Dath was, obviously, not one's usual tedious contract work. Dath's writing as a fiction author, as a music critic – serving as chief editor for *SPEX*, the most influential German magazine for popular culture, between 1998 and 2000 – as an interdisciplinarily ambitious and unconventional but erudite Marxist theorist and as an experimental essayist let him appear in hindsight as an almost congenial choice. Dath embodies in his writing most of the styles and skills and areas of knowledge and critique that also Eshun embodies – with all the differences in the intellectual life in Germany or the UK at the time. The easiest passages to translate were therefore those that attached to his reading experiences and also his own writing style:

> All the passages (I don't have them in my head now, but there were quite a few) that reminded me a bit of the New Wave of science fiction from the sixties/seventies (the 'New Worlds'-sound, Moorcock, Ballard, etc.) in style and choice of words were very fast, that's the tone of voice I grew up with myself

as a science fiction reader, also regarding a certain tone in the corresponding German translations. (Dath 2018b;[1] translated by Holger Schulze)

More difficult were apparently some aspects referring to free jazz in *More Brilliant than the Sun*. Though Dath is an encyclopaedic listener and expert in a wide array of genres and performance histories, obviously, certain missing links come only to one's attention when actually encountering other listeners – in this case Kodwo Eshun – who energetically and full of excitement point at this very musician or musical genre:

Several things in connection with (Free) Jazz I had to get my head around; I didn't want to germanize these passages blindly, i.e. after imagining how something would probably sound that K.E. [Kodwo Eshun] writes about, and so I took a kind of crash course in these things, Alice Coltrane especially, I hardly knew at all, I benefitted from this greatly – and I only knew clichéd stuff about Sun Ra, which I hope I did develop further into a better understanding by listening more closely. (Dath 2018b;[2] translated by Holger Schulze)

Dath's Mixillogics

Heller Als Die Sonne was published by ID Verlag, a leftist and experimentalist publisher from Berlin that has focused since the late 1980s on giving 'the homeless autonomous and militant left a publicistic mouthpiece' (Knoblauch 2017). This choice of publisher was truly fitting in comparison to the original publishing house of Quartet Books. ID Verlag published research on and around the history and the theory of antifascist movements, collected writings of anti-imperialist and revolutionary groups since the 1970s such as the Revolutionäre Zellen/Rote Zora in Germany or the Weather Underground in the United States – and much later also writings and theories on black electronic music. Eshun's book on electronic music, afrofuturism and revolutionary approaches towards sound cultural research therefore blends perfectly into this programme. The original book, though, could be advertised successfully to the

English-speaking community of readers just by the household name of its author, at that time already a prolific music critic, essayist and intellectual figure in the UK; yet, this was not so much the case in Germany, at least not in the year 1999 the translated book was being published. In this publishing context, a shocking new approach to sound, to electronic music, and to writing about electronic and largely afrocentric music and sound simply had to convince its readers through other means: be it through its dedicated first readers, reviewers, journalists, propagandists and cultural disseminators in general, be it through actual resonance in cultural and academic institutions and their discourses, be it through word of mouth from artists, readers and, not least, from fans of the original publication. This translation also profited from the mythical and allegedly widespread success of the publication already in the original language.

Translating a monster of a text such as *More Brilliant than the Sun* is basically an almost impossible and, hence, largely poetic task. In this case, though, in respect to Eshun's quite impressive writing style, Dath had on top the difficult task not only to recreate the book's argument in another language (German) – but also to introduce, to recreate and to regenerate the author's numerous neologisms, puns, portmanteaus, and even his rhapsodic flow, grown out of years of working as a music critic, in this new language. This task is then not merely poetic, it resembles more a sort of co-authorship with time delay. Translating this book might have been at times, I can only imagine, as difficult as a translation combining *Infinite Jest* with the *Grammatologie*, the *Xenogenesis* trilogy and *Finnegans Wake*. Dath's own background as a science fiction reader and novelist apparently helped him a lot in doing so. Around translating Eshun he wrote and published over fifteen novels with a dystopian, utopian, an alternate history if not a decidedly science fiction setting underneath – beginning in 1995 with *Cordula killt Dich! oder Wir sind doch nicht Nemesis von jedem Pfeifenheini. Roman der Auferstehung* (Cordula Kills You! or We Are Not The Nemesis Of Every Pipe Dreamer. A Novel Of Resurrection, 1995), over the award-winning *Die Abschaffung der Arten* (2008; translated 2013 as *The Abolition of Species*) to the most recent *Der Schnitt durch die Sonne* (The Cut Through The Sun, 2017). His music and concert reviews, his essays on musical aesthetics have discussed a broad variety of genres between dance pop, metal, hard rock and hip hop (Dath 2007: 81–124). He was and is known for newspaper articles that bordered again more on the genre of experimental essayism than on the genre of reportage or political analysis – leading him to publications on Marxist theory or an introduction to the writings of Karl Marx (Dath 2018a). Dath's writing is interweaving arguments, figures of thought, cases and exemplifications from critical and Marxist theory as well as an always surprising line of concepts and terminologies from sub-disciplines and research areas in mathematics, the natural or the engineering sciences, with rather corporeal, often intimate and highly suggestive narrations of a situated and sensory experience – combining a sort of analytical high tone with a set of distinctly profane and vernacular idioms, activities and observations. With *Heller Als Die Sonne* (Eshun 1999) he most markedly moved into the area of music criticism and sound studies. If one takes a closer look at his music writings one can see how his style is replete with rhetoric figures, style characteristics and figures of thought from Eshun's approach of sonic fiction.

Take this review of Madonna's tenth studio album *Confessions on a Dance Floor* from 2005, written and published by Dath seven years after *Heller Als Die Sonne*. In this review – again in the *Frankfurter Allgemeine Zeitung* – Dath actually narrates the sonic experience and fictions related to, oozing out, or just vaguely associated to the songs, the production and the biography of this pop persona's latest record at the time. Under the title *Sie malt die Nacht mit Licht an* (She paints the night with light), he begins his review with a poetic account of popular culture as an intergalactical sacrifice to higher entities – positioning Madonna, the artist, as questioning this tradition:

> The smartest producers of modern times have always celebrated what the Swedes left to mankind as a feast of lightness and grace, as something pure, holy, a greasy wedding noodle floating far from space. Madonna, however, exposes for 'Hung up' the other, the dirty and demanding, in short: the brutal side of the 'Abba'-experience, the heavy tracked vehicle of love, the high-tech dance tank. (Dath 2005b;[3] translated by Holger Schulze)

From this beginning, Dath writes his way through the record thereby connecting descriptions of its musical production techniques with again more mythical accounts of its sensory effects and affective values, its *articulations of need* (Koppe) – densely filled with metaphors and allusions:

> And here's how it goes on, at a consistently high level: 'Get Together' sounds as if it has been programmed under water by thinking bathing essences on atomic submarine navigation computers, 'Sorry' fetches ancient basses from the cellar of the pyramids and shoots them at the clouds, 'Future Lovers' juggles acoustic magnetic fields and paints the night with stroboscopic light, 'I Love New York' builds a sounding city of rhythmically sorted hot flushes between steep concrete walls – it's all about synaesthetic things, says this story. Images and fragrances are always included in this. (Dath 2005b;[4] translated by Holger Schulze)

In these selected paragraphs from a 1,000-word review, Dath writes with and through all three constituents of sonic fiction: *mythscience*, *mixillogic* and *mutantextures*. In this case, though, they are not employed to discuss or to illuminate cultural artefacts from afrofuturism but one cultural artefact from the sphere of dance pop: a translation that can seem surprising, maybe even inappropriate, but that actually excavates even here, on occasion of a major commodity of pop culture, its connecting traces to more remote areas of vernacular culture. When Dath refers to 'what the Swedes left to mankind as a feast of lightness and grace, as something pure, holy, a greasy wedding noodle floating far from space' or to 'being programmed under water by thinking bathing essences on atomic submarine navigation computers', 'fetches ancient basses from the cellar of the pyramids and shoots them at the clouds' he insinuates a seemingly inconceivable, transdimensional *mythscience* of pop production; when Dath lists the combinatorics of 'synaesthetic things' and then concludes 'images and fragrances are always included in this', he recognizes and iterates the radical deviating credo of *mixillogic* in these productions; and when he then describes 'the heavy tracked vehicle of love, the high-tech dance tank', a song that 'juggles acoustic magnetic fields and paints the night with stroboscopic light' and another one that 'builds a sounding city of rhythmically sorted hot flushes between steep concrete walls' then these colourful images represent vividly the tangible and audible *mutantextures* woven into these songs. It is apparent that in this sonic writing any explicit afrofuturist and black diasporic tie is almost totally lost and erased; only in its inclination towards afrocentric, aquatic and afrofuturist imagery the original context from Eshun's invention of sonic fiction is retained. However, in a more benevolent if not mixillogic reading one can surely assert that writing about elaborately evolved production techniques in a German conservative newspaper of the 2000s still can and must be considered at the time a partly alien if not diasporic endeavour. It is, actually, a confrontation with sensibilities and technologies that the author performs in this review.

His understanding and his employment of radical and drastic aesthetics regarding technologies and sensibilities Dath illuminates most clearly in his most concise and outspoken poetics *Die salzweißen Augen. Vierzehn Briefe über Drastik und Deutlichkeit* (The Salt-White Eyes. Fourteen Letters About Drastics and Directness, 2005); there he quotes the film studies scholar Linda Badley:

> The fantastic is based in somatic consciousness – in sensational existence that is tragically conscious of its material finitude and the presence of Otherness, in the torture, challenge, and horror-comedy of incessant change. (Badley 1995: 35).

Even in his writing about a global pop persona such as Madonna, Dath confronts with his fantastic essayism indeed a *somatic consciousness* with *the presence of Otherness* and *of incessant change* in style and in subjects. This very articulation of need – 'the dirty and demanding, in short: the brutal side of the "Abba" experience, the heavy tracked vehicle of love, the high-tech dance tank' (Dath 2005b) – generates its mutantextures out of a mixillogic of senses and techniques. As a consequence, Dath's writing fabricates through rhetoric mixillogics similar mutantextures that can also be found in Eshun's writing. Both authors' styles of writing never present a neatly organized unfolding of a carefully prearranged and meticulously dissected and deliberately balanced argument. However, precisely this more erratic and wilfully distracted writing strategy led Dath then to work on a large volume of social theory, written together in a seemingly more academic manner with chemist, writer, and long-time collaborator Barbara Kirchner: *Der Implex. Sozialer Forschritt: Geschichte und Idee* (The Implex. Social Progress: History and Idea, 2012). This concept of the implex and its discussion by Dath and Kirchner then bears a surprising relation to and an insightful interpretation of sonic fiction.

The Dialectics of the Implex

In 2012 Dietmar Dath published, together with Barbara Kirchner, a large volume of over 800 densely set pages that bears a title which might for most of its readers seem strange, alien, if not totally unintelligible: *Der Implex*. In the German as well as in the English language the word *implex* is not a household term. Already reading it aloud qualifies as an encounter of the third kind with an alien breed of vocabulary none of us might have encountered before. *The implex*: what could that be? What could this mean? Is it an alien *doomsday machine*? Is it a kind of hyperdimensional *vortex*? Or more some excessively powerful superintelligence that is – in the most radical sense of the word – intangible and inconceivable by mortal beings such as you or me who will most probably never leave the precincts of this planet? Moreover, in what kind of discourses and transdisciplinary conversations does one enter when engaging in a discussion of the implex? In the preface of *Der Implex*, Kirchner and Dath, the author duo, answer this question:

> The key question is whether something like social progress can be thought and, more importantly, made. One could say that this book is a kind of fiction in concepts: it accompanies the fates of attempts to make the world a better place than people of the modern age found it when they began to be people of the modern age. (Dath & Kirchner 2012: 15;[5] translated by Holger Schulze)

The mutantexture of this book is, hence, a *fiction in concepts*, a *Roman in Begriffen* (in the German original). By assuming this perspective the authors' whole endeavour is then situated in the midst of this hybrid genre and writing style of theory-fiction or concept fiction. Further they write:

> As in any historical novel, love occurs here as well. But the hero of the book is a concept that we found in Paul Valéry and then enriched and changed for purposes other than his: the implex.

What it means to us is not explained at length, but is shown in the scenes mentioned, in the wild and in action. (Dath & Kirchner 2012: 15;[6] translated by Holger Schulze)

The *scenes* of this concept fiction are the historically documented *attempts to make the world a better place*, made by modern nations in their sciences and economies, their epistemologies and arts, their ethics and their warfare (Dath & Kirchner 2012: 15). In the eighteen sections of this long treatise – consisting of five to twelve chapters each – the authors scrutinize the histories of Marxist and other projects of social transformation in a sort of inspired time travelling. They review in their analysis the potential and the actual effects on everyday life of these projects – all in the light of the concept of the implex.

The implex as a figure of thought, however, is not as murky as its sound might suggest. Basically, it is a dialectical concept that refers to an ongoing process of deep and intrinsically predetermined transformations. This concept gives a name to the well-known constellation of inherent intentions and potential trajectories of actions that are implied in a given situation, a given person or group of persons, in an institution. Valéry's original concept focused solely on intrapersonal implexes, describing sensibilities, inclinations and the implied intentions ruling and unravelling them – an intricate theory of sensibilities, if you will, that will be discussed in the next section of this chapter. The interpretation by Dath and Kirchner, however, transfers this concept from the area of the personal to the area of the social. In both cases the term implex describes a goal that was not explicitly conceivable or even tangible at an earlier stage. Yet, once the differing later state has been reached and can be conceived explicitly, precisely this earlier and implied stage is then being called the implicit goal, the implied complex of intentions, or *the implex* of the earlier one. In the simple but precise and thoroughly German bureaucratic words of Dath and Kirchner:

> bestimmte nicht unwahrscheinliche Folgelagen seien der Implex einer spezifische Ausgangslage gewesen. (Dath & Kirchner 2012: 44)

But if *certain, not improbable subsequent situations were the implex to a specific starting situation,* then also a sort of, if you will, forward engineering could be possible. One could ask – in line with some of the utopian scenarios of the twentieth century: *what will have been the social and political, the economic and technological implex of this current situation?* Yet, such an analysis requires, as the authors state, a sort of *Implexaufmerksamkeit*, a *sensibility for implexes*. These dialectical relations between a set of material prerequisites and subsequent social or political developments are especially poignant in Dietmar Dath's essay on Karl Marx's theory:

> For Marx, capitalism is a historically singular incident in which a form of exploitation produces so much wealth that the abolition of exploitation can be put on the agenda. If one does not see the existing false situation as simply a mistake that goes astray due to false ideas, but as the only available reservoir for the right practice then one will rather make fun of people who believe it would be enough to exorcise the false ideas. (Dath 2018a: 54–55;[7] translated by Holger Schulze)

In this Marxist interpretation the implex is materially accessible only in the expanded constellation of all artefacts of a society, its economy, its sciences and publication media, its technologies and its art forms. These artefacts and practices altogether almost coerce if not protrude an implied transformation into a yet unimaginable – but later on rather consequential – state of this society. Such a description can easily sound like a magic trick that would be able to turn any set of technological inventions and new commodities into surprisingly progressive social developments. However, Dath and Kirchner emphasize in their argument the volatility and the indeterminacy of such a social progress. There is no determinism, neither in Marxist theory of the steps towards communism nor in Dath's and Kirchner's interpretation of the implex of liberation movements finding their prerequisites in new technologies. They exemplify this by referring to the washing machine, the dishwasher and their potential function of overturning the heteronormative power structure of the bourgeois family and its gender roles:

> The washing machine or the dishwasher have knocked a few weapons out of misogyny, however, this was nowhere and never sufficient for corresponding social changes; this detailed observation already contains everything one should know about

the chances of any further elimination of the division of labour as a breeding ground for hierarchies, exploitative conditions, exclusion, etc. (Dath & Kirchner 2012: 808;[8] translated by Holger Schulze)

Capitalism and its exploitative conditions are, following Dath and Kirchner, neither fully determined by simply a false consciousness nor just a given material or technological set of circumstances alone. In order to prove this, they introduce, in good materialist tradition, an experience from everyday life and domestic work as their major example. This example materializes then vividly the indeterminate character of social and historical developments. Nevertheless, an existing *sensibility of implexes* and the ability to imagine a differing society – maybe as a *fiction in concepts?* – is a necessary constituent and a potential motor of social progress. This is the complex dialectics of the implex: from hindsight it might seem almost too simple to analyse, historically, the crucial constituents of a social change that soon after took place in this specific social and historical situation; yet, if one experiences this very situation as present times, it is not trivial and obvious at all that this very set of new technological innovations might bear the potential to transform the given social and political institutions into something yet completely unimaginable. The implex is a volatile and malleable quality in society: it desperately needs the action, the activism, the intervention, also the protest and the critical, the vital, and the truly innovative and revolutionary energy of many protagonists. Only these actors on the stage of politics and social protest can indeed, following Dath and Kirchner, materialize some still imaginary constellation of an implex into actual social progress. *Eine Bedürfnisartikulation verwandelt sich in politisches Handeln: an articulation of need turns into political activism.* The implex is real, it is material and existing – yet it remains a mere potential if no one cares to actualize it with one's own energy and life and political agency.

Valéry's Sensibilities

The origin of the concept of the implex can be found mainly in several brief passages scattered all over the work of French essayist and poet Paul Valéry. In his famously erratic *Cahiers*, the notebooks he wrote all his life, beginning in 1894, he developed this concept and he explicated it in various other works outside of these notebooks. In *Idée Fixe*, a so-called 'A Dialogue at the Seaside' first published in 1932, Valéry lets one protagonist say:

> The implex…is [our] ability to feel, react, do, understand – individual, variable, more or less perceived by us – and always imperfectly, and indirectly (like the sensation of fatigue), – and often misleading. (Valéry 1965: 56)

In this dialogue, the concept of the implex is presented as a neologism to describe the inherent sense for something, the directed and vectorial energy in all the sensibilities present in a person. It is significant that this concept is introduced in a dialogue between *He* and *I*, between *a doctor* and *Monsieur Teste* – so, actually between two character traits or personae of Paul Valéry himself (Burghart 2013: 243–248). In this self-reflection in the mode of an externalized and staged dialogue Valéry explores how the scientific knowledge and concept of a human being, of man and of self had changed recently in the twentieth century – and how this might or might not have affected one's actual self-reflection. Eight years later then, he explicated a bit further how an implex could represent a person in all its ambivalences, complexities and dynamics. In 1940 he notes, under the moniker *Sensibility* in his *Cahiers*:

> Implex, is basically what is implied in the notion of person or self, and is not of the present moment. It's the potential of general and specialized sensibility – of which the present is always a matter of chance. And this potential is conscious. (Valéry 2007: 221)

Valéry, therefore, claims that the potential and all the future actualizations that you or I might perform or act out, are present in you or me in a sort of, as one could say, complex and implicit way – in one's sensibility. This potential is not explicit, it is not yet clear to what end it might lead, but it already seems to point in a variety of directions, it contains *vectors*, so to speak. However, these vectors might (or might not) be realized in the near or far future the way one could imagine them being realized. In the end, this realization only fulfils an implicit goal, this actual telos of the implex – of which neither you nor I might have any idea right now what it could be one day. This concept of the implex resembles very closely Robert Musil's concept of the *sense of possibility* or *Möglichkeitssinn* (Musil [1930] 1978: 16–18; Márquez 1991; Bauer & Stockhammer 2000). Both concepts bear the birthmark of a shaking ground in philosophy and epistemology around 1900 – which apparently led their inventors to provide specific figures of thoughts to speak about *a potential that might (or might not) be realized.* Yet Musil's concept of a sense of possibility is more focused on very specific actions to be taken (or not), decisions to be made (or not) and events to be triggered (or not) in direction of alternate or possible worlds; whereas Valéry's concept of the implex includes also specific actions, decisions and events in direction of future developments, but these are then always understood as constituents of a much larger tendency, a constellation of sensibilities in one's individual (not collective) life that might (or might not) be realized: Musil's sense for potential actions to create possible worlds differs in this respect vastly from Valéry's reflection on one's sensibilities and how they can potentially unfold into possible actions and activities. These individual sensibilities are of no major interest for Musil who cares more for a kind of almost objectivist overview of varying timelines and alternate histories that might be developing out of certain actions, decisions and events. For Valéry, though, precisely these sensibilities are the indulgently subjectivist material and the medium out of which the implex is made. An implex embodies for Valéry a complexly implied constellation of sensibilities, idiosyncrasies, maybe obsessions and fears, desires and irritations that circulate or linger in our persona. It is a phenomenological and introspective concept that intends to explicate of what kind of material all these more distinct decisions in our lives are formed and made. Valéry would claim that they emerge exactly out of these vague and blurry clouds and constantly malleable, often unclear inclinations, desires, scepticisms that linger in our persona, in one's soma. This presence of a vague yet present prerequisite for future decisions and developments is not actually focused on by Musil – but it provides the framework for the theories explicated by Dietmar Dath and Barbara Kirchner in their book on the implex.

Dath and Kirchner now take this concept of intrapersonal transformations of sensibilities, transfer it into a more sociological and historical discourse and expand it into a new nucleus for a general theory of social progress. Whereas Valéry's reflections are solely rooted in an individual's – actually, in Valéry's – personal reflections and intentions, this is not so much the case in Dath's and Kirchner's application. Their inquiry under the title *Der Implex* is a book that tries to investigate the dynamics, the obstacles and the successful strategies when struggling for social progress. They try to understand what makes social and political change at all possible – and what are some of the prerequisites that help to distinguish potentially futile endeavours from more promising and hopeful ones. This basically revolutionary interpretation of the implex differs therefore massively from Valéry's introspective understanding. Still, there is one major link between both approaches that must not be overlooked: Valéry considers the reflection on personal implexes a general, maybe even a societal task for the sciences and for research – and in the same way also Dath and Kirchner consider their general inquiry on social progress actually rooted in rather personal intentions, proclivities and trajectories. These comparable elements in both approaches are nevertheless dwarfed when one delves into the details of Dath's and Kirchner's ambitions.

In contrast to Valéry's various notes and statements, both authors expand and apply his concept not only to formal logic, to genealogy and to poetics (Dath & Kirchner 2012: 44) – but they mutate it, as discussed in the previous section, to become a convincing marxist political concept; with this they go then much further than Derrida in his discussion of the implex (Derrida 1972). Their political concept starts out with the truly communist intention to transform societies and their societal strata on a political level. Transforming society though necessarily requires and often also implies certain constellations that provide surprising scientific discoveries and insights. Only these scientific discoveries can then in turn pave the way for inventions effectively driving these crucial transformations that might lead to a revolution. Political transformations are therefore – following Dath and Kirchner – equally implied in scientific discoveries as in social transformations (Dath & Kirchner 2012: 42).

This is exemplified in *Der Implex* by the main example of the Industrial Revolution and its inventions in the nineteenth century:

a revolution in the double sense that on the one hand provided the means for an accelerated capitalization and exploitation of workers and underclasses – but at the same time it also provided the means for new and more powerful forms of workers' associations than ever before. The Industrial Revolution promoted factually the political revolution – a genuinely dialectic and Marxist *Denkfigur*. As a consequence, Dath and Kirchner also assume that contemporary transformations regarding globalization and digitalization might have similar dialectical effects: the revolution of computerization, digitization, automation and globalization might promote in the end an even more substantial political revolution than ever before. The implex is at play in all these cases.

The implex of a situation is therefore defined as an inclination towards a certain direction of further development or action, implying – if not demanding – a collective or individual generativity. Cautiously though, Dath and Kirchner negate all teleological or even eschatological necessity in this process: it is still required to actually respond to and to deal with the many coincidentalities affecting it. It is, I would like to repeat at this point, not a deterministic approach – but it proposes a more generative, transformative and versatile understanding of societies and cultural developments. One could even claim, in turn, that such an idiosyncratic implex constitutes the distinct core of all dialectical and generative approaches that do not strive for a reduction of all humanoid aliens at all times and on all areas of this planet to a supposed, static common denominator and anthropological constant, under all circumstances and mutations. The approach of the implex – as I can find it also in Eshun's concept of the sonic fiction – accentuates to the contrary the fundamental malleability and the non-linear development, the cultural and sensory potential and affectability of those aliens at least who are living and roaming and failing on earth. For Dath and Kirchner this generativity and intentionality in an implex is not limited to one person alone – transcending here Valéry's original concept – but to a whole econo-cultural and socio political constellation in all its stupendous complexity and transversal implications. Afrofuturism, sinofuturism, xenofeminism or queer futurism and the many other specimens of ethnofuturisms actually follow along this path.

Even Wrong Ideas Can Be Made True

In a small introduction to the writings of Karl Marx, Dath focuses on a pragmatic and even relativist turn that primarily pragmatists such as Charles Sanders Peirce and William James performed in the nineteenth century but that one can detect apparently also in Marx's thinking. Dath writes about Marx's materialist pragmatism:

> A thing, he says, exists only if you can do something with it, and only then if you have an accurate imagination of that thing, can you do what you want to do based on that imagination. (Dath 2018a: 52;[9] translated by Holger Schulze)

Is Dath here interpreting Marx with Land's concept of theory-fiction in his mind? Fabricating a fiction that seems to be so real that it actually can have direct effects in real life as it provokes people to take action? In one section of his Marx book with the title *Even Wrong Ideas Can Be Made True* (*Selbst falsche Ideen kann man wahr machen*) Dath is then indeed explicating Marx's inventive intellectual energy by an effectively pragmatist goal: the goal to provide useful tools for thinking and imagining, for a vital revolutionary effort that actually would lead to a series of untamable uprisings and riots, amounting to a revolution. This drive to materialize revolutionary ideas did then eventually produce,

> An encyclopedia of historical possibilities, realized and missed; of liberation movements, their material prerequisites and the reasons for their failure; a compendium of theories, both unused and expired. A dialectic lesson reflecting on progress, an insisting on reason in history – which is not a ladder, but an at least four-dimensional, undirected ensemble of possibilities and situations. An arsenal of sharpened instruments of critique: critique of ideologies, of comfortable thinking, of not thinking at all. (Dath & Greffrath 2018:[10] 38; translated by Holger Schulze)

Fictional possible worlds, also sonic ones, are in this understanding already transforming and promoting social progress. They propose and inspire a multiplicity of options and activities – the immensely rich contingency of differing, of diffracting histories, of processes of social negotiation, of political decisions and the many specimens of cultural heritage. The concept of the implex underlines these inherent tendencies in a given societal and historical, cultural and biographical constellation – and it accelerates their development towards an aspired differing state in the near or far future: maybe only achieved after a series of individual or collective actions, mutations, falsifications, revisions or amplifications? Sonic and sensory fictions and their implex motivate and inspire actors to engage in social practices and social interpenetrations to alter society. These issues regarding how to achieve social progress resonate therefore directly with the main drives of afrofuturism and the various other resulting futurisms, be it sinofuturism or Shanghai futurism. The concept of futurism in these efforts and their interpretations and appropriations is factually close to identical with a societal transformation through technopoetics, particularly *black technopoetics*, that can be traced back to the ancient and somewhat prehistoric Italian futurists to which Chude-Sokei pays respect:

Its importance is worthy of note if only for being the first futurism, without which Afro-futurism, astrofuturism, queer futurism, Chicana-futurism, Kongo-futurism, and others – would suffer for want of a suffix. (2016: 12)

The early historical futurisms of the twentieth century, though, were then nurtured by European culture wars and fuelled by the energy of an overheated art market at the time; the contemporary or more recent specimens of futurisms ask differing questions, shaped more by public discourses on liberation and on anticapitalist critique:

How does the future meet us halfway? How can we think freedom and emancipation beyond any antiquated logic of progress? In other words, how can we envision a political horizon beyond the hegemonic traditions of historicism that still inform the political realities of Europe or North America—and, consequently, much of the rest of the world too? How can we develop the ability to produce a history or deny historical fabrications differently from traditional Western culture, not least in its explicitly colonial and racist tendencies? In what ways can all of us who think about the possible implications of concepts such as progress or emancipation today include in our thoughts and agendas the political subject of the twenty-first century: the refugee? (Avanessian & Moalemi 2018: 8;[11] original English version by the authors)

Technopoetics, or – to be more precise – *sociopoetics*, the generative practices of societal progress in the twenty-first century ask coherently one major question, again and again: *how can we develop the ability to produce a history or deny historical fabrications differently from traditional Western culture, not least in its explicitly colonial and racist tendencies?*

3

Black Aurality
Alien Sonic Nontologies

They visit you. You do not know where they come from. They deport
you. They take you and your families and your friends, your kids
out of the habitat in which you and your ancestors have lived now
for years, decades, centuries if not millennia. Then they ship you –
days and weeks and months without perspective, with no hope
of returning at some point to your home, to your elders, to your
friends and families – into some radically unknown new territory.
The vast void you have been shipped over might at some point even
have become your second home: a home in forced migration, in
deportation containers. Yet, now you are here, on a new world. This
is truly an alien territory to you, where you are now and of which
you know absolutely nothing. You have no clue where you actually
are, what you are supposed to do here, how you are expected to
behave, and what awaits you at the end of this enforced deportation
and, somehow, incarceration on alien territory. You are – figuratively
and literally at the same time – somewhere far, far away; surely
in another galaxy, in another, alien dimension. However, you are
now regarded as the alien here. You have been deported to this
world you will probably never leave again – you will never be
allowed to leave again. You are at the mercy of those who brought
you here. "The New Mutants are the outcasts" (Eshun 2005: 218).

In writings, in movies and in songs, in music videos and stage
performances, in sleeve notes and in aesthetic reflections this
brief narration is being encircled, extrapolated and executed in
afrodiasporic thinking and culture. It is the, if you will, crucial

anti-origin experience at the core of afrofuturism (Dery 1994; Akomfrah 1996; Steinskog 2018). This experience of deportation and of alienation is the starting point for afrofuturist art and design, for its music and theory, and its practice – but not necessarily in a tone of defeat or resignation; but, to the contrary, with a vivid and dynamic energy of reinvention, of radical rebellion, and even of a kind of superiority – often somewhat concealed – radiating from this core experience. Those who were forced to live all their lives as aliens have one significant benefit above all those who never were deported from their habitat: they surely had to experience a violent extraction, an extradition from all the ties and lies and false consciousness, the whole ideological character of naturalized identities. Naturalist or even essentialist illusions are hard to believe for them. In the words of the collective xenofeminist author by the name of Laboria Cuboniks:

> We are all alienated – but have we ever been otherwise? (Laboria Cuboniks 2018: 15)

From this core experience of afrofuturism – and more recently various other forms of futurisms, such as 'Xenofeminism, Sinofuturism, Dubaification or Gulf Futurism' (Laboria Cuboniks 2015; Avanessian & Moalemi 2018), 'astrofuturism, queer futurism, Chicana-futurism, Kongo-futurism, and others' (Chude-Sokei 2016: 12) – emerges a surprising energy, a seemingly unstoppable urge, a continuous prolific generativity, a productively alienated generativity that invents and founds and constructs a whole new continuum of historiographies, of epistemologies, of ontologies that bear next to no similarities to the ones established by academia in traditional disciplines and their recognized forms of knowledge. These are alien epistemologies, alien ontologies, alien historiographies. Again, in the words of Laboria Cuboniks:

> The construction of freedom involves not less but more alienation; alienation is the labour of freedom's construction. (Laboria Cuboniks 2018: 15)

Alienation is at the core of afrofuturism. It is a driving force behind sonic fiction, behind all sorts of sensory and theory-fiction. It is a constructive and an epistemological force, a force towards social change and progress. All these more recent futurisms are not *l'art pour l'art* projects residing on an imaginary island of erratic retrofuturist renewals as they are sometimes portrayed. Actually, all of them are connecting to the implex idea of social change as expressed by Dath and Kirchner. They are,

> Fiercely insisting on the possibility of large-scale social change for all of our alien kin. (Laboria Cuboniks 2018: 44)

This urge for social change is then, necessarily, in conflict with existing institutions, dispositives and power structures: as soon as one indeed conceptualizes new alien sensibilities, alien auralities, black auralities and ontologies. Especially ontologies and the recent turn to them, also in *sonic materialism* has been critically scrutinized regarding some of their racialized and essentialist examples (a substantial and well argued critique of xenofeminism on that basis can be found in Goh 2019). Most prominent is probably the more recent critique by Marie Thompson of sonic materialism:

> The ontological, meanwhile is naturalized as universal ground, obscuring the realm of non-being upon which it is predicated. Thus where the ontological has come to signify 'a realm of apparent liberation from the miasmas of the social world' in much realist and new materialist thought, Fanon regards ontology itself as 'a mystifying form of appearance that posits itself as outside of social inscriptions of race, when in fact this very positing is integral to the dialectics of racialization itself.' (2017: 268)

Thompson points here convincingly at the universalist exclusion mechanism of ontologies that function as highly implicit and a priori. With reference to Fred Moten and Franz Fanon she shows how such a universalist use of ontologies is fundamentally an operation of territorialization and colonization – and as such it already represents a factually racist and non-inclusive 'white aurality':

> It amplifies the materiality of 'sound itself' while muffling its sociality; it amplifies Eurological sound art and, in the process, muffles other sonic practices; it amplifies dualisms of nature/culture, matter/meaning, real/representation, sound art/music

and muffles boundary work; all the while invizibilizing its own constitutive presence in hearing the ontological conditions of sound-itself. (Thompson 2017: 274)

However, as Fred Moten writes: 'The history of blackness is testament to the fact that objects can and do resist' (Moten 2003: 1) – and afrofuturism, sonic fiction and black aurality are examples of precisely this resistance:

> This ability to talk back – to simultaneously celebrate in sound and offer philosophical intervention, to critique – is crucial as we develop different strategies to negotiate our ethical and political lives. (Havis 2009: 757)

Consequentially, also a Black Aurality is marked by its specific 'histories, practices, ontologies, epistemologies and technologies of sound, music and audition' (Thompson 2017: 274), its specific material-discursive composites. Yet, in the case of 'whiteness and aurality [the] material-discursive composites that shape and are shaped by one another and in relation to a particular environment' (Thompson 2017: 274) are more often concealed. This *particular environment* and its *material-discursive composites* are habitually covered up, necessarily and shamefully, as otherwise the inherent violence, the crime, the immoral and the inhuman ongoing practices of colonialism would be overly present in every single moment a person born in a colonial nation would raise its voice. This holds also true for me, being the author of these lines, who disclosed earlier at least some of his biographical traits; but still, the white aurality I was raised and educated in is, apparently, even now shaping my efforts of respectfully and supportively making the case for alternate and black auralities. Yet, only at a much, much later, maybe imaginary point in history, when white aurality might not any longer be considered an unquestionable and objective approach to listening and sounding, only then this effort of *decolonizing the aural* would have proven successful. For now, truly both – and many more – specimens of aurality need to be materialized in their excessively idiosyncratic historical, societal, practical, ontological, epistemological and technological specificities. No specimen of aurality at all can be rightfully regarded as an unmarked and absolute 'ahistorical, unchanging perceptual schema' (Thompson

2017: 274). All auralities might – sadly so – not have been actually created equal. It is therefore a researcher's task to grant them an equally detailed and intense attention to question, to explore and to scrutinize their constituents and their generative traits.

Black Aurality

Black aurality can be found in some of the first crucial writings about the Black Atlantic – the main area that deportation ships crossed as part of the slave trade organized by European colonial empires and industries. This space of transition has since then been considered one nucleus for the artistic, literary, musical and research practices around afrofuturism. In 1993 Paul Gilroy defined and described the Black Atlantic as follows:

> The specificity of the modern political and cultural formation I want to call the Black Atlantic can be defined, on one level, through this desire to transcend both the structures of the nation state and the constraints of ethnicity and national particularity. These desires are relevant to understanding political organizing and cultural criticism. They have always sat uneasily alongside the strategic choices forced on black movements and individuals embedded in national and political cultures and nation-states in America, the Caribbean, and Europe. (1993: 19)

Gilroy rejects the territorial notion of eurocentric nationalism and expands the notion of home to the actual non-territory of the ocean, the Atlantic, that factually provided the major environment for the slave trade, the colonial commerce and the ongoing general traffic between Europe, Africa and the Americas. Therefore, houses or family trees, farming ground or material soil do not become the foundational structure for his approach but the very instruments of forced mobility and migration. Motion and movement are the main forms of activity and the main figures of thought for descendants from the Black Atlantic:

> Ships immediately focus attention on the middle passage, on the various projects for redemptive return to an African homeland,

on the circulation of ideas and activists as well as the movement
of key cultural and political artifacts: tracts, books, gramophone
records, and choirs. (Gilroy 1993: 4)

Tracts, books, gramophone records, and choirs: the constituents
of an afrofuturist and a Black Atlantic identity that Gilroy lists
here are various historical media formats of performed, recorded
and inscribed articulations that could indeed travel on ships at the
historical times in question. They were small and portable enough
to be transported between harbour cities and continents – yet
they were also capable and versatile enough to indeed carry valid
messages, artistic performances and cultural representations. It
is those media formats, apt for travelling and cultural exchange,
that Gilroy proposes as constituting the core of the Black Atlantic
circulation of artefacts. These formats and artefacts were escorting
and also supporting if not nobilitating the people being transported
over the Atlantic against their will: being deported into alien worlds.

From this starting point in the *middle passage* and in the
circulation of ideas and activists and artefacts black aurality needs
to be conceptualized. But before more closely discussing *black*
aurality – what is *aurality* anyway? The concept of aurality as such
is first of all, and maybe against a reader's intuition, not identical
to the concept of sound culture. This concept would be defined as
*sets of listening and sounding practices linked to sets of sounding
and listening apparatuses*. Aurality, to the contrary, has a broader,
a more abstract, and a much more general scope. The aurality
of a historical period or a specific cultural area (as explicated
for example by Erlmann 2010 or Gautier 2014) implies and
defines not necessarily only a specific set of material cultural
practices of listening and sounding and their apparatuses
themselves. Aurality represents the general approach of a culture
towards the auditory senses and sonic sensibilities for the whole
of this culture. It does not solely apply to its sound culture in the
narrowest sense but to all aspects of the economy, of administration
and governance, of finance and investment, of the arts and of design,
of the sciences and the humanities, of crafts and housekeeping, of
entertaining and of everyday life. It structures by its approach to
the aural and the sonic the culture pervasively. In the narrowest
definition though, aurality can refer to the 'shared hearing of

written texts' (Coleman 2007: 68) as the most common document of historical and cultural knowledge: it then means the listening practices directed at literature or language-related performances. Aurality, in this established sense, means the role that listening and the aural has in a given culture (cf. Coleman 1996). So, whereas a specific *sound culture* is to be excavated from the actual practices, the material culture, and the apparatuses dominant in a culture or a historical period, the *aurality* of a culture or a historical period represents a more pervasive, underlying and structuring constituent that might not even result in actual sound practices or listening apparatuses. The aurality of a culture refers to its main assumptions, its knowledge, and its ontological, epistemological and anthropological insights and positions regarding listening and sounding – at times even confirmed by and discussed on occasion of actual sound practices and listening experiences, but not necessarily so. A *black aurality* therefore is defined by a distinct set of such assumptions, forms of knowledge as well as ontological, epistemological and anthropological insights and positions regarding listening and sounding. The role and insights into aurality represented in afrofuturism – for example in the writings by Gilroy, Delany, Dery, Butler or Eshun, in the performances and the musical works by Sun Ra, George Clinton, King Tubby, Dr Octagon, Drexciya or Janelle Monáe – embody and perform such a distinct set of those assumptions and positions.

A black aurality of this kind operates, understandably, on a quite different and differing programme than the locally established and therein nobilitated articulations of white aurality: an aurality that is also foundational for the culture I am now writing and publishing this book in. White aurality is, referring again back to Marie Thompson's reflections, inextricably tied to an economic, social and political system of exploitation, of slavery, apartheid and capitalization; black aurality, in contrast, was historically for a more recent time period operating on the receiving end of the activities of white aurality – and from this experience also on the side of an energetic and resisting reinvention of its guiding concepts for listening and sounding. Black aurality would be the aurality of the actors, entities, objects, also the subjected beings that had no choice but to be objects of the mistreatment by a patriarchal, capitalist and colonialist white aurality; as an articulation of resistance and subversion – black aurality produces therefore a harshly differing specimen of sounding and listening, of living with sounds and of performing sound, of being performed by sounds and of reinventing all of this, continuously. Performing an enforced dynamization and mobilization, a continuous transport is one of the main characteristics of this aurality. The most provocative difference here is precisely articulated by Stefano Harney and Fred Moten:

> The ordinary fugue and fugitive run of the language lab, black phonography's brutally experimental venue. Paraontological totality is in the making. Present and unmade in presence, blackness is an instrument in the making. Quasi una fantasia in its paralegal swerve, its mad-worked braid, the imagination produces nothing but exsense in the hold.... Blackness is the site where absolute nothingness and the world of things converge. Blackness is fantasy in the hold. (2013: 94–95)

This fugitive imagination and *fantasy in the hold* of which Harney and Moten speak, this generative imagination – *an instrument in the making* – of differing, of diffracting and generative new ontologies, this motion is constitutive. Black aurality cannot be separated from black fugitivity, a historical and present reality in black culture that Tina M. Campt interprets, following Harney and Moten, as the logical consequence of consistent alienation, deportation, criminalization and rejection by a society structured through white supremacy:

> It's the refusal to be a subject to a law that refuses to recognize you. It's defined not by opposition or necessarily resistance, but instead a refusal of the very premises that have historically negated the lived experience of Blackness as either pathological or exceptional to the logic of white supremacy. (Campt 2014: 47:38–48:20)

This 'refusal to be refused' (Harney and Moten 2013: 96) is generative also in its effort to note, not to conceal, but to transform, if not to transcend the 'sonic color line' (Stoever 2016) that factually represents the border drawn by white aurality to limit the motion and everyday transgression of unbounded lives. This sonic colour line embodies the racialized listening to persons refused by white supremacy; it is a cultural driving force behind colonial deportation and a prerequisite for the contemporary existence of black aurality. A black fugitivity in sound is then, historically, being performed as a sonic subversion:

> Slavery was a most vicious system, and those who endured and survived it a tough people, but it was not…a state of absolute repression. A slave was, to the extent that he was a musician, one who realized himself in the world of sound. For the art – the blues, the spirituals, the jazz, the dance – was what we had in place of freedom. Techniques (i.e. the ability to be nimble, to change the joke and slip the yoke) was then, as today, the key to creative freedom, but before this came a will toward expression…enslaved and politically weak men successfully impos[ed] their values upon a powerful society through song and dance. (Ellison 1995: 856)

The sonic and sensory colonialism that bound those performers and artists in the first place and deported them into this alien nation represented a structural societal force that relied, after all, heavily on the *audiovisual litany* as detected by Jonathan Sterne (2012: 9):

> Indebted to the spiritualism and the ascendancy of the white Christian West, the audiovisual litany is organized around a series of dualisms that treat visual and sonic experience as unchanging and transhistorical givens. (Thompson 2017: 271)

This fallacy of the *transhistorical given* is then what is mainly attacked by black aurality, by afrofuturism and by employing sonic fiction. They all reshape, they reorder and rethink, they bend and deform the truisms and so-called eternal truths of a territorializing white aurality and its sonically transcendentalist legitimation. As soon as one recognizes the limitations and incarcerations of hegemonial white or vanilla auralities, then the urge for a differing, a progressive approach to auralities is the consequence. But how could one reasonably generate such an aurality differing from the one being dominant these days?

The Diffraction of a Mythscience

A differing aurality, differing from a hegemonic one, incorporates a *diffracted* reading. The concept of *diffraction* in academic practices of research has been introduced by Karen Barad as a critical concept for *feminist materialism* (2003, 2007, 2011, 2012). She proposes – in reference to the works of Donna Haraway and of Trinh Minh-ha on the one side and on the other side to the physical theories of optics – the optic function of *diffraction* as an alternative to the well-known optic function (and implicitly also to the metaphor of intellectual activity) of *reflection* (Barad 2007: 71–96). The goal of diffraction in epistemology now, as proposed by Barad, is to incorporate the factually existing differences and differentials relevant for one concrete research process – without eliminating and concealing them. Its goal is therefore, contrary to an often performed synthesis after laying out before all the critical and self-contradictory elements, not a final assimilation and repression of distinct differences after reinstating their difference. To the contrary, with diffraction one performs a careful unfolding of the many minuscule diffracting constituents given in order to get an insight into how these diffractions affected one specific research process – but also into the diffracting qualities of the specific methods and operations one is performing with, the tools one employs when working as a researcher.

This approach then can be applied to materials, to academic texts, and to all 'entangled practices [that] require[s] a non-additive approach that is attentive to the intra-action of multiple apparatuses of bodily production' (Barad 2007: 94). With this approach in mind, any process of research, but also other generative processes of everyday life can be analysed in regard of the manifold diffracting forces constituting crucial transformations. It neglects the radical and full intentionality of actions in humanoid aliens; it also rejects the notion of a pure, true, radical or untainted approach or reading; and, to the contrary, it accepts the material outside and all the pervasive forces that shape your and my actions all the time. This form of critique proposes for the course of a discussion process not to invent or to extract radical oppositions between which one then would need to position oneself or stage a fateful decision. With the concept of diffraction, instead, it becomes possible to accept that a series of diffracting agents are ubiquitously and incessantly present in any process of action – it is much more promising to analyse the intricate qualities they adjoin or subduct in every single case from the process of action, be it in research or in other areas of society. It is an analytical practice of respect and of subtle differentiations:

> Reading diffractively therefore not only appears to transcend the level of critique, ultimately based in a Self/Other identity politics, but in Barad's regard also can be regarded as a boundary-crossing, trans/disciplinary methodology, as it brings about 'respectful engagements with different disciplinary practices.' (Geerts and Tuin 2016, quoted in Barad 2007: 93)

As a consequence the concept of 'diffraction allows you to study both the nature of the apparatus and also the object' (Barad 2007: 50). This concept now can be applied to the field of sonic fiction in general and of afrofuturism in particular especially in regard to both their relations to a hegemonic interpretation and historiography. White historiographies and sciences habitually need to neglect and to repress the impact of any diffractions or distractions, any not so marginal influences or outside, material forces on their actions; this ignorance regarding their own intrinsically heteronomous genealogy is a core practice of their territorializing and colonialist approach of white supremacy. There simply cannot be any other. Yet, this excessively self-indulgent ignorance is next to impossible for black historiographies or black sciences, mythsciences. They are basically generated by at least one major diffraction if not disruption. In this respect, the whole discursive practices, the musical, the literary and the academic works of afrofuturism are in themselves complexly layered examples for a thoroughly diffractive methodology. Usually, though, diffraction is applied to readings of academic texts and the canon as well as academic experimental settings and research methods – but it can also be applied to a whole set of cultural texts and their canon. This is what afrofuturism precisely does. It does not and cannot possibly claim there is no connection to the white narrations and the white ontologies and epistemologies out of which its very own material entanglements emerged. But it proposes a distinctively bent, a deformed, a substantially diffracted reading. An afrocentric and afrofuturist world is obviously diffracted from a eurocentric world – yet it still is entangled with all the artefacts, texts, the cultural productions, and the aesthetics of the eurocentric world it diffracted from. A more recent definition of afrofuturism than the famous one by Mark Dery from 1994 (cited in the chapter 'What is Sonic Fiction?' in this book) precisely states this diffracting character:

> Afrofuturism can be broadly defined as 'African American voices' with 'other stories to tell about culture, technology and things to come.' The term was chosen as the best umbrella for...'sci-fi imagery, futurist themes, and technological innovation in the African diaspora.' (Nelson 2002: 9)

This definition clearly lays out the deforming and mythscientific operation inherent to afrofuturism as being a deviant narration and articulation of *culture, technology and things to come*. To a large degree this diffracting mythscience then is founded on its sound cultures and on black aurality, which is why it received soon thereafter also the name of *Sonic Afromodernity* from Alexander Weheliye (2005). Understood as a diffractive reading of history, of modernity and of sound culture, sonic afromodernity and afrofuturism precisely 'do not care about canonical readings of texts or of artefacts because they zoom in on how texts, artefacts and human subjects interpellate or affect each other' (Tuin 2018: 101). This diffractive practice becomes especially performative in artistic and literary readings and explorations of afrofuturism; famously explored by Kodwo Eshun for example in the mutantextural recordings of Drexciya and their diffracting mythscience:

> Each Drexciya EP - from '92's *Deep Sea Dweller*, through *Bubble Metropolis, Molecular Enhancement, Aquatic Invasion, The Unknown Aquazone, The Journey Home* and *Return of Drexciya* to '97's *Uncharted* – militarizes Parliament's 70s and Hendrix's 60s Atlantean aquatopias. Their underwater paradise is hydroterritorialized into a geopolitical subcontinent mapped through cartographic track titles: *Positron Island, Danger Bay, The Red Hills of Lardossa, The Basalt Zone 4. 977Z, The Invisible City, Dead Man's Reef, Vampire Island, Neon Falls, Bubble Metropolis*. The Bermuda Triangle becomes a basstation from which wavejumper commandos and the 'dreaded Drexciya stingray and barracuda battalions' launch their Aquatic Invasion

against the AudioVisual Programmers. Every Drexciya EP navigates the depths of the Black Atlantic, the submerged worlds populated by Drexciyans, Lardossans, Darthouven Fish Men and Mutant Gillmen. (1998: 83)

This passage from *More Brilliant than the Sun* now could be read as just a superficial play with associations and references, the suggestive sound of certain terminologies and place names; however, it represents an actual and material diffraction from the official historiography of the slave trade into an only slightly alternate, slightly diffracted world as performed in the liner notes to the album *The Quest* by Drexciya:

Could it be possible for humans to breathe underwater? A foetus in its mother's womb is certainly alive in an aquatic environment. During the greatest holocaust the world has ever known, pregnant America-bound African slaves were thrown overboard by the thousands during labour for being sick and disruptive cargo. Is it possible that they could have given birth at sea to babies that never needed air? Recent experiments have shown mice able to breathe liquid oxygen. Even more shocking and conclusive was a recent instance of a premature infant saved from certain death by breathing liquid oxygen through its undeveloped lungs. These facts combined with reported sightings of Gillmen and swamp monsters in the coastal swamps of the South-Eastern United States make the slave trade theory startlingly feasible. Are Drexciyans water breathing, aquatically mutated descendants of those unfortunate victims of human greed? have they been spared by God to teach us or terrorise us? Did they migrate from the Gulf of Mexico to the Mississippi river basin and on to the great lakes of Michigan? (Drexciya 1997: liner notes)

This sonic fiction by Drexciya gets then expanded and elaborated; it gets interconnected to and infused by other discourses and aesthetic traditions in Eshun's book: instead of putting diffraction to a halt and reducing it analytically, Eshun effectively expands and accelerates even the ongoing process of diffraction, into a dynamized form of *syrrhesis* (which will be discussed in the following chapter). With a comparable strategy, Thomas Meinecke, German musician, radio host and author, further expands precisely this diffractive fictionalization, primarily in a book from 2001: in *Hellblau* – translated by Daniel Bowles as *Pale Blue* (Meinecke 2012) – the remarkable colour that constitutes the title represents such a diffracting blurriness across all areas of life. While exploring origin histories, relationships, demarcations and interpellations between vinyl records, theoretical treatises, fashion items, dancing experiences, intimate relations and sexual practices between the protagonists (Meinecke 2001, 2012), the author and the personae populating his writings actually encounter forms of uncertainty, ambivalence and an observation of all the diffractions present. These diffractions are constitutive ambivalences that help to elaborate all the details inherent to black aurality and mythscience:

> RuPaul says: I am black, I am gay, I am a man, and I love being all these things. RuPaul, as genetically male Ultrafeminine. RuPaul's peroxide-blond wig on his shaved head. RuPaul as black blonde. RuPaul's Back to My Roots video, in which he presents annoying blond Afro hairdos to the general racist gaze. (Meinecke 2012: 5; 37)

Gender roles, narrations of heritage and notions of blackness or whiteness, national languages and production styles of electronic music are put into diffraction and continuous deconstruction and reconstruction in Meinecke's writing: these diffractions simply never end, they are instead respectfully observed and read – in all their intricate entanglements, mutual affectations and interpenetrations. To a point that this diffraction is also performed corporeally and sonically in the material and the corpus of writing, in the rhetorics of diffracting repetitions and insistent vortices – also when speaking, digesting and joyfully tasting the name of particular performers, musicians and groups such as Dopplereffekt or Drexciya. In the words of Eshun:

> The name Drexciya is an adventure for the tongue. You hold a geography in your mouth. 'Drex': the tongue descends a staircase, ascends on 'ci', skips on 'ya'. The sublime tastes good to speak. (1998: 126)

Diffracting mythsciences begin with mixillogic sensations of this sort; they expand then into observable deviant practices – and they do not yet end with bodily experiences and corporeal forms of knowledge that matter mutantexturally. *You hold a geography in your mouth. The sublime tastes good to speak.*

Alter Nation, AlterDestiny and Autohistoria

'Tell me, do you know how to use a sonic cleaning plate? That's what I've got in the back.' – 'No.' ... She gave a little laugh. 'You don't? ... Well, do you at least know how to use a damned squat-john? All I need is to have you pissing and shitting all over this hulk like it was your putrid rat cage.' (Delany 1984: 20–21)

This world is alien. The world that Samuel R. Delany narrates here, in his novel *Stars in My Pocket Like Grains of Sand* from 1984 is very different on all sorts of levels: starting with gender denominations, family structures, caste systems, sexual practices, professional activities and forms of transport. In this quote it is only the rather marginal example of a so-called *sonic cleaning plate* – that apparently serves as a kind of showering or bathing facility to clean one's body. But the protagonist – who recently agreed to his own enslavement due to his character and intrinsic desires, not for external reasons – seemingly is not familiar with this seemingly very common technology; and so aren't we, the readers. This world is unfamiliar in so many aspects that it is hard to find a point where to start with exploring, let alone understanding it.

Afrofuturist writings, compositions and artefacts represent – as in this example – a set of cultural productions that transcend the framework of existing and white epistemologies, white historiographies and white ontologies: they enter an alien continuum. Only to mark the differing set of ontologies present therein as *black* is therefore not at all sufficient. It would merely reverse the existing order to an alternate structure that still would adhere and mirror the present one. In the words of Sun Ra:

I speak of different kind of Blackness, the kind
That the world does not know, the kind that the
world
Will never understand

It is rhythm against rhythm in kind dispersion
It is harmony against harmony in endless
coordination
It is melody against melody in vital
enlightenment
And something else and more
A living spirit gives a quickening thought.

                                                    (Sun Ra 2005: 295)

'*The kind that the world does not know, the kind that the world will never understand*': these sentences not only articulate a set of 'countermythologies' (Eshun 1998: 158). The desire for an alternative destiny or 'AlterDestiny' (Sun Ra; cf. Langguth 2010) articulated here aims actually at a sort of *nontology*, an *autohistory*. These articulations of need transcend immensely the more common and easily attainable desires. They represent a radical cut with maybe still condoned threads to the hegemonic white ontologies. Yet, the alienation began long ago:

The ships landed long ago: they already laid waste whole societies, abducted and genetically altered whole swathes of citizenry, imposed without surcease their values. Africa and America – and so by extension Europe and Asia – are already in their various ways Alien Nation. No return to normal is possible: what "normal" is there to return to? (Sinker 1992: 33)

Alienation is not a process yet to happen. It is a historical prerequisite for the state we are in now. This state already is an *Alien Nation*: 'You are the alien you are looking for' (Eshun 1998: 84). But where to go from here, from this state of dispossession and appositionality? Moten and Harney ask:

Can this being together in homelessness, this interplay of the refusal of what has been refused, this undercommon appositionality, be a place from which emerges neither self-consciousness nor knowledge of the other but an improvisation that proceeds from somewhere on the other side of an unasked question? Not simply to be among his own; but to be among his own in dispossession, to be among the ones who cannot own, the ones who have nothing and who, in having nothing, have everything. (2013: 96)

With these questions of a self-inquiry a substantial process of transforming *black fugitivity* (Moten 2003) into a *black futurity* (Campt 2017), into an alternate history and destiny might begin:

> Technology generates the process Sun Ra terms an AlterDestiny, a bifurcation in time. The magnetron migrates across the mediascape, changing scale from Marvel Comics 60s supervillain Magneto, leader of the Evil Mutants, to Drexciya's Intensified Magnetron, to Killah Priest's 'magnetron which puts your arteries back apart.' (Eshun 1998: 85)

They are *black technopoetics* (Chude-Sokei 2016) that generate an AlterDestiny by diffraction, a materialized bifurcation in time. This AlterDestiny through technopoetics being propelled into black futurity is diffraction in action: this action is a process of *cultural decolonization* (Mignolo 2011) through a process of technopoetic expansion and of materialized fictionalization. Its starting ground in threatening racializations though cannot be lost as Ayesha Hameed reminds us:

> Sun Ra's project can only make sense in the wake of racialised slavery in America and the genocide during the middle passage. It is not a whimsical flight of fancy but rather a structured protest whose flight is inextricable to the violence that it is responding to. (Gunkel, Hameed, O'Sullivan 2017: 9)

Sonic fiction is a proposal of how to skilfully craft and arrive at an AlterDestiny. With this craft it is an inextricable and fundamental constituent of the infinite task of cultural decolonization. In the words of sound design researcher and decolonization thinker Pedro Oliveira this process proceeds by a,

> Decolonization of knowledge, spirit and the self, exactly by seeing them as inextricably related, an entanglement of bodily knowledge, political identity, and ancestral reconciliation. (Olivera 2017: 42)

Afrofuturism is then, consequentially, the framing cultural practice that energizes these subversive as well as revolutionary activities of decolonizing African cultures – and can promote a decolonization also of all the other colonized cultures on this globe. This process then includes also an experience of a libido and a love that accepts to be rooted in this diasporic alienation:

> Afrofuturist love, then, is a love that paradoxically yet strategically remakes alienation as Alien Nation. (Veen 2016: 86)

From this state of an Alien Nation though, a diffracting practice of narration and self-narration can start as well. For instance, conceptualized as the *Autohistoria* that Gloria Evangelina Anzaldúa proposes as part of a decolonizing process (cf. Oliveira 2017: 42–43) – in her case not to advance black futurity but, as one could argue, *Mestiza Futurity* or *Sensory Mestiza Fiction*:

> Autohistoria is a term I use to describe the genre of writing about one's personal and collective history using fictive elements, a sort of fictionalized autobiography or memoir; an autohistoria-teoría is a personal essay that theorizes. (Anzaldúa 2002: 578)

An autohistoria is therefore a sensory essay as personal fiction: a fiction that also expands the life of persons in an Alien Nation who seek an AlterDestiny into a realm of futurity. This fiction is self-reflective, it is aware of its genuine diffraction and proceeds nevertheless in precisely this fictionalizing and reflective way. This reflective methodology is then diffractive:

> I call a diffractive methodology, a method of diffractively reading insights through one another, building new insights, and attentively and carefully reading for differences that matter in their fine details, together with the recognition that there intrinsic to this analysis is an ethics that is not predicated on externality but rather entanglement. (Barad 2012: 50)

Diffractive in the case of autohistoria are the *code-switching*, the *queer epistemology* and the *border culture*: the *code-switching* (Anzaldúa [1986] 2009) between several languages or language-like codes in everyday life, for example 'Standard English, working class and slang English, Standard Spanish, Standard Mexican Spanish, North Mexican Spanish dialect, Chicano Spanish ... Tex-Mex, Pachuco' (Anzaldúa [1987] 2012: 55), between modalities of writing or drawing, between narrating vague memories of dreams or documented encounters during daytime; the *queer or chicana feminist epistemology* (Calderón et al. 2012; Dahms 2012) materializes 'the changeability of racial, gender, sexual, and other categories' and ratifies the disruption of the 'binaries of colored/ white, female/male, mind/body' (Anzaldúa, 2002: 541); and the all-encompassing *border culture* is then 'constructing a hybrid text that moves between different types of written expression' (Lockhart 2006), juxtaposing a poem with a discoursive account, colloquial memoirs with a more strictly academic discussion, overheard songs with a finely crafted avant-garde text. In the words of Andrea J. Pitts, autohistoria is:

> Collaborative, sensuously embodied, and productive of critical self-reflection, which can be both harmful and enabling. (2016: 357)

These diffractive qualities bring it as close to sonic fiction as can be. Autohistoria incorporates mixillogics of code-switching, mythsciences of queer epistemologies and the mutantextures of border culture. By conceiving, writing and establishing an autohistoria even the state of Alien Nation can be moved further towards a potential AlterDestiny. A diffracted history, epistemology and ontology – like in the small quote from Delany's *Stars in My Pocket Like Grains of Sand* – it bends the common notions of all these white disciplines into increasing areas of alterity. Even into the unsettling and intransigent realm of the *NON* (that will be explored in Chapter 6); even into the mythscience of Sun Ra:

> It hardly matters whether the story's true or figurative, hallucination or bad neural wiring, that's the point where the Jazzman breaks away from the standard riff and makes up his own melody. Here, in his front room, all cluttered up with disciples' pictures of himself as Egyptian deity, as cosmic explorer, as mystic messenger, he tells the ordinary story of an ordinary abduction by aliens and then – because he is Le Son'y Ra, and not as other corny tale-spinners – he tells how he turned down the offer of Messiahship. (Sinker 1992: 30)

Decolontologies

Jes Grew has no end and no beginning. It even precedes that little ball that exploded 1000000000s of years ago and led to what we are now. Jes Grew may even have caused the ball to explode. We will miss it for a while but it will come back, and when it returns we will see that it never left. You see, life will never end; there is really no end to life, if anything goes it will be death. Jes Grew is life. They comfortably share a single horse like 2 knights. They will try to depress Jes Grew but it will only spring back and prosper. We will make our own future Text. A future generation of young artists will accomplish this. If the Daughters of the Eastern Star can do it, so can they. What do you say we all go down to the restaurant and have a sandwich? (Reed 1972: 204)

The prolific energy of afrofuturism, its diffracting negation generated and still generates a multitude of alternate historiographies – not only the mestiza culture or border culture just mentioned in the section before. Recently, this still expanding multiplicity has been summed up as a series of *ethnofuturisms*, by Armen Avanessian and Mahan Moalemi:

The notion of a black secret technology allows Afrofuturism to reach a point of speculative acceleration. ◊ Blaccelerationism proposes that accelerationism always already exists in the territory of blackness, whether it knows it or not. ◊ Sinofuturism is a darkside cartography of the turbulent rise of East Asia; it connects seemingly heterogeneous elements onto the topology of planetary capitalism. ◊ Shanghai futurism ultimately depends on breaking free from the now common assumption about the nature of time. ◊ The unfolding story of Gulf Futurism is a strange mitosis happening out of the sight of the masterplanners and architects; it's the splitting of worlds, of then and later, us and them, real and unreal. ◊ The Dubaification of the world is already a thing of the present and the recent past, and has completed its ideological mission at lightning speed. (Avanessian & Moalemi 2018: 7;[1] original English version by the authors)

The diffracting generativity of these new futurisms, these new *nontologies and autohistories* is seemingly endless and unstoppable. They will expand and prevail – even more so as the colonialist and imperialist imprints on the US-centric production of afrofuturist fiction, theories and music is more and more pointed out, also by Kodwo Eshun recently:

> Many contemporary artists and critics within the continent object to the perceived Americocentricity of Afrofuturism. They argue that Afrofuturism fails to account for the preoccupations that inform practices produced in the past and the present throughout the cities of the continent and the Caribbean. In Johannesburg, Nairobi, Lagos and Accra, novelists, theorists, bloggers, photographers and filmmakers are beating 'the planetary turn to the african predicament', which Achille Mbembe argues 'will constitute the main cultural and philosophical event of the twenty-first century.' (Gunkel, Hameed & O'Sullivan 2017: 265)

In this direction a growing amount of research in *Black Sound Studies* (Nyong'o 2014; Chude-Sokei 2016: 167; Steinskog 2018: 1–36) also inspires research in more of these *ethnofuturist* areas:

> Ru Paul says: Who says black people have to be black. (Meinecke 2012: 5)

One might recognize here a *nontology* to overcome existing narrations, epistemologies and ontologies: a multitude of ethnofuturist ontologies to end narrations of phylogenetic or ontogenetic progress. These nontologies are then *decolontologies*: they decolonize and autohistorize the territorialized and racialized areas of ontologies. Sonic fictions contribute to and materialize these decolontologies.

> Humanity in its attempt to destroy itself had made the world unlivable. She had been certain she would die even though she had survived the bombing without a scratch. She had considered her survival a misfortune – a promise of a more lingering death. And now…?

'Is there anything left on Earth?' she whispered. 'Anything alive, I mean.'

'Oh, yes. Time and our efforts have been restoring it.'

That stopped her. She managed to look at him for a moment without being distracted by the slowly writhing tentacles. 'Restoring it? Why?'

'For use. You'll go back there eventually.'

'You'll send me back? And the other humans?'

'Yes.'

'Why?'

'That you will come to understand little by little.'

(Butler 1997: 15)

4

Sensory Epistemologies
Syrrhesis and Sensibility

You stand in sound. Right now. Right here. Wherever you are. It engulfs you, it envelops you, it pinches and cuts through you. It is everywhere you are, too. With your limbs and sensibilities, your longings and repulsions, with your hopes and your indolences you sense and you react to these sounds, you project or trigger certain other sounds.

Kodwo Eshun's book *More Brilliant than the Sun* is not a book written about music. *More Brilliant than the Sun* is a book written *out of* music. This sentence makes almost no sense in traditional epistemology and neither does it in a contemporary framework of commodified research in a peer review culture of the early twenty-first century. Eshun's approach of writing about sounds, sound culture, technocultural traditions as deviant afrofuturist nontologies and autohistories though is far from being a neat and scholarly unfolding of propositions and arguments. He never introduces his readers to sonic fiction nor afrofuturism. He throws you, the reader, into a swirl out of all of these epistemes and artefacts, of percepts and experiences, imaginations and technologies and many more. Yet, being thrown into all of this grants an intense experience to the reader and listener, sensor and reflector in this sonic thinking:

> You are not censors but sensors, not aesthetes but kinaesthetes. You are sensationalists. You are the newest mutants incubated in wombspeakers. (Eshun 1998: -001)

This experiential rhetoric transforms one's position as a reader. I can feel the bass hit my intestines. I enjoy that. When the groove drags along my muscles and bones. When the beats drive into my limbs and muscles, nerves and feet. From usually being the evaluating and scrutinizing reader of academic books or essays one transforms into the experiencing and imagining body of a reader of novels and fiction and poetry. As a reader of fiction one does not necessarily intend to evaluate mainly the quality of arguments, their inconsistencies or logical fallacies, the terminology they carry with them, the underlying assumptions. To the contrary, when reading fiction one might be more inclined to let oneself be guided by the author, following all the imaginations carried with any pleasantly sounding sequence of language-like words, radiating rhythms and references and imaginations: let the sonic traces of the words sink into your mind. To put it bluntly: in *academic reading* the reader reviews and evaluates critically the quality of the author's writing – in *fictional reading,* though, the reader appreciates and accepts, in general, an author taking control of the reader's imagination. The reading situation in itself is a completely different one.

The inversion of control when reading, though, does not imply that these two modes of reading and of control are strictly separated along the lines of academic non-fiction and fiction for the mass market. Fictional passages, interjections, arabesques and erratic detours are genuine means of stylistic freedom for writers of non-fiction; and, similarly, instructive and educative passages, interjections and reflective detours are also continually used in fiction. The radical element in Eshun's writing is not the selective use of unconventional stylistic means: that would just confirm its stylistic coherence in general. No, Eshun performs in his writing a continual and thorough reversal of fundamental assumptions about academic or non-fiction writing. Namely, that such a writing is fundamentally rooted in the model of the proof, of the pleadings or the jurisdictional argument in court; to the contrary, Eshun's *sonic writing* (Kapchan 2017; Schulze 2019b) incorporates the generativity of sound in academic writing:

As soon as you realise that sound/audio space/acoustic space, however you define it, has a generative principle – that it is cosmogenetic in a sense and that it can generate its own world picture – you're off. Then the technical machine isn't just a technical machine, it's a vector out into the world. (Eshun in Weelden 1999)

A well-trained routine of academic writing is then mutating into a generator of possible worlds, not only sonic possible but *sensorially possible worlds* (Voegelin 2014). With this approach then Eshun connects to the wider field of non-fiction writers who make narrative, experiential, sensory and experimental forms of writing an integral part of their articles, essays and monographs. Indeed, the most prominent and adventurous or even influential writers and scholars applied certain sensory and narrative strategies of fiction in their essays (Stanitzek 2011; Dillon 2017). One might start only in the twentieth century with Walter Benjamin's writings on his *Berliner Kindheit um 1900* (*Berlin Childhood Around 1900*, 1950), meet halfway at Roland Barthes's *Mythologies* (1957), not finding an end in Audre Lorde's *Sister Outsider* (1984), Peter Handke's *Versuch über die Müdigkeit* (*Essay on Tiredness*, 1989) Gloria Anzaldua's *Borderlands/La Frontera: The New Mestiza* (1987) or the film-maker Alexander Kluge's *Chronik der Gefühle* (*Chronicle of Emotions*, 2000), Maggie Nelson's *Bluets* (2009) or *Too Much and Not the Mood* (2017) by Durga Chew-Bose. Or one might even get back to the early and genre-defining essays by Michel de Montaigne and later to the erratic rhapsodies, aphorisms and notes by Nietzsche or Georges Bataille, but also look at the idiosyncratic genre hybrids written by Hunter S. Thompson or Joan Didion – or look at all the examples of recent autofiction between Catherine Millet, Annie Ernaux or Karl Ove Knausgård (Gasparini 2008; Grell 2014; Dix 2018). None of these authors write fiction. But all of them make use of narrative, of poetic, of suggestive and even experimental writing strategies that evoke and enliven their object of reflection. They manage to seduce a reader to escort further the train of thoughts, the entangling self-reflections and the enveloping affects and sensory experiences of an author. They are *theory fictions* in the original, if you will, *pre-Landian* sense. They do *not intend* to seamlessly brainwash you into some ideology at hand – though, nevertheless, one might find, obviously, ideologies in there, ensheathed with the sweet skills of essayistic writing. They actually intend and achieve to open up, often anxiously and doubtful their process of reflection and hesitation, thinking and sensing by using imagination and narration, poetry and experiments in writing. 'I am hesitant, says the third tongue' (Serres 2008: 163).

The premodern prehistory of this generative non-fiction style is long: in European history the *Essais* written by Michel de Montaigne in the sixteenth century are taken as a starting point with all the moralists, aphorists and non-academic thinkers and writers along the way as stepping stones towards a modern history of the essay. Montaigne's reflections and explorations then gave the whole genre of non-fictional but still personal, highly sensible if not intimate writing its appropriate name. These writings are essays, *il's essaient*, they essay and try, they probe to think about their topics, they might fail and try again, then fail again and fail maybe, at some point, in the far future, a bit better. This effort and the growing public interest in it is apparently an effect that only plays out in subsequent centuries with the emergence and the rise to power of the male bourgeois subject in the eighteenth century – replacing in the then new European nation states and republics the aristocratic, oligarchic and patriarchal order with a democratic but still patriarchal and largely plutocratic order. Nevertheless, with all its crimes and self-blinding arrogance, with colonial warfare, exploitation and torture, with white supremacy and capitalist desires at its core, this transformation still represents, sad to say, one of the more noble cultural achievements of the world region I am born in. A self-reflective and self-questioning writing of the twenty-first century though – in a period when the bourgeois subject can in no respect any longer be regarded as the sole entity of authorship – other literary strategies, rhetoric positions and hybrid genres, other literary performativities and textual personae develop this genre into a postcolonial, intensely mediatized, networked, into an intersectionally aware, gender-reflective and yet economically and technologically accelerated world.

The main question of an essay, maybe being written 500 years ago, is though still valid: how can new forms of knowledge, new insights, new propositions come – under these altered conditions – into the world? How can one excavate such new insights? And is it even thinkable that – besides the experimental sciences, empirical studies and deductive or critical arguments – also the arts, the design, and even personal, maybe intimate practices of everyday life could contribute to these new insights?

The Body of the Researcher

In 1985 a book was published that was neither a novel nor a theoretical treatize in the traditional sense (Serres 1985). It began with the words:

> Fire is dangerous on a ship, it drives you out. It burns, stings, bites, crackles, stinks, dazzles, and quickly springs up everywhere, incandescent, to remain in control. A damaged hull is less perilous; damaged vessels have been known to return to port, full of sea water up to their deadworks. Ships are made to love water, inside or out, but they abhor fire, especially when their holds are full of torpedoes and shells. A good sailor has to be a reasonable fireman. (Serres 2008: 17)

This is not the start of an academic study. This is the beginning of a poetic novel, possibly with a strong inclination towards experimental and suggestively immersive forms of performative writing, now and then rather in an erratic style. Being a trained seaman, Michel Serres, the author of *The Five Senses*, this book and this beginning, takes his readers into common practices of fire training on a marine ship. Material, corporeal and sensory practices incite the author's reflection:

> Fire training demands more of the sailor and is harsher and more uncompromising than anything that he needs to learn as a seaman. I can still remember several torturous exercises which teach not only a certain relationship to the senses, but also how to live or survive. We were made to climb down dark, vertical wells, descending endless ladders and inching along damp crawlways, to low underground rooms in which a sheet of oil would be burning. We had to stay there for a long time, lying beneath the acrid smoke, our noses touching the ground, completely still so as not to disturb the thick cloud hanging over us. We had to leave slowly and deliberately when our name was called so as not to choke our neighbour with an ill-considered gesture that would have brought the smoke eddies lower. (Serres 2008: 17)

As readers we are taken – no: we are forced right into all the sensory and experiential aspects of fire training, of time structures of urgency, of bodily postures, of ache and of tense moments of waiting. This is not an abstract reflection, no armchair philosophy, no self-indulging *Glasperlenspiel*. No author intends here primarily to expand his or her publication list, citation index or bibliometric impact. The common 'lazy, pompous, lard-arsed, top-down dominance' (Eshun 1998: -004) of scholarly mannerisms and strategic research branding is hardly to be found here. This is an everday life's practice in its most suggestive, intense and truly epistemologically insightful version:

> The breathable space lies in a thin layer at ground level and remains stable for quite a long period. Knowing how to hold your breath, to estimate the distance to the heart of the blaze or to the point beyond which one is in mortal danger; how to estimate the time remaining, to walk, to move in the right direction, blind, to try not to yield to the universal god of panic, to proceed cautiously towards the desperately desired opening; these are things I know about the body. (Serres 2008: 17)

Michel Serres's knowledge and skills as a writer, thinker and researcher radiate exactly from this example of corporeal knowledge. It does not find its start in science history or the history of philosophy or of theories of perception. It starts on a ship, burning, hurtful, between dangerous water, threatening fire and a precisely shaped social and pragmatic situation. It starts with tasks and goals, rituals and regulations, material and sensory realities, with potentials and constraints. *'These are things I know about the body'*. He writes:

> This is no fable. No-one sees dancing shadows on the walls of the cave when a fire is burning inside. (Serres 2008: 17)

In other words: there is no cave. There is no Platonic *Allegory of the Cave*. And, consequentially, there is also no history of philosophy, epistemology or ontology as such – at least these histories seem a bit less urgently pressing and relevant when a life-threatening fire burns close to you and you need to react, to do something, to move on and to flee. It is *carnal sociology* (Wacquant 2015). It is corporeal and sensory thinking. All armchairs of philosophy turn to ashes then, following Michel Serres. Or in the words of Kodwo Eshun:

> There is no distance with volume, you're swallowed up by sound. There's no room, you can't be ironic if you're being swallowed by volume, and volume is overwhelming you. It's impossible to stay ironic, so all the implications of postmodernism go out of the window. Not only is it the literary that's useless, all traditional theory is pointless. All that works is the sonic plus the machine that you're building. So you can bring back any of these particular theoretical tools if you like, but they better work. And the way you can test them out is to actually play the records. (1998: 189)

Both Eshun and Serres claim alike that there is no distance with volume and heat and intensity and presence. You are swallowed up by sound, smell, physicality, haptic kinaesthetics, vertigo, entanglement, desire and affect. There is no room: you cannot possibly be ironic if all of these are swallowing you. All *traditional* (read: representational, anthropocentric, disentangled, distanced and armchair-happy) theory, so they write and show, becomes pointless. So, what is then left to do? If traditional theory won't work – maybe some *non-traditional* theory could work? You might still be able to invent new forms of thinking and conceptualizing that are maybe more appropriate to present situations of experiential and practical entanglement. As Eshun writes:

> All that works is the sonic plus the machine that you're building. So you can bring back any of these particular theoretical tools if you like, but they better work. And the way you can test them out is to actually play the records. (1998: 189)

Translated into Serres's anthropology of the senses this means: all you can do as a thinker, writer and researcher under such demanding circumstances is to start anew from these moments of visceral and sonic intensity. You start right here and right now with your sensibility and your body as a researcher – and you build from here a new kind of constellation of thought figures, of concepts, and explicating models, of epistemological trajectories circulating around these experiences you made, as a researcher. *These are things I know about the body*: a new *theory-machine*. One might then even concede at some point: you can of course bring back certain of those theoretical tools you had dismissed earlier, if you really like – but only under one condition: *they'd fucking better work*. The question that arises then is: how can theoretical tools actually *work* – aside from the common colloquial phrase 'that doesn't work for me'? Obviously, a working theoretical tool is neither a case for elaborate labour theory nor for a mechanic calculation of work in physics, including energy and power. Theories *work* primarily in the sense that they are capable of expanding on, of broadening one's perspective on, maybe even of punctually explicating how, and with what implicit goal, in what manifold relations and under what conditions a certain observed and experienced phenomenon of reality takes place the way it does. One can then *work with this theory*. I can think *with this theory*. A theory *works* therefore as a generative nucleus and as an accelerating thought figure in thinking and writing.

As a consequence, Eshun demands from a theoretical tool a final test – that most theory might not pass. He proposes to test the quality of a theory precisely under those situated circumstances, those moments of tense pressure, momentary intensity, affective sensibilities, and rapid action sequences that are actually characteristic for the practice that a theory is trying to explicate: 'And the way you can test them out is to actually play the records' (Eshun 1998: 189). Therefore, in Eshun's – and I would add also in Serres's – *sensory epistemology* any insight of scholarly writing needs to pass this test. For both authors it is simply not enough to write about fire or water, sound or touch, taste or smell when ignoring the actual material circumstances and visceral effects of smell and taste, touch and sound, water or fire. Letters, words and propositions on a book page are simply not enough. *These are things I know about the body*. They demand to be used in this situation they reflect upon. This harsh *reality check* breaks then with some more traditions of science history, epistemology and research. It neglects in the end also the division of labour in research. Because, if research and its results would indeed be that strictly detached and separated from the areas they claim to explicate and to interpret, they were, following Eshun and Serres, factually rendered useless. This leads Serres then to a deconstruction of yet another platonic core text: the *Symposion*.

In one of the core chapters of *The Five Senses* – under the title of 'Animal Spirits' – he explores the sensory, gustatory, olfactory and metabolist qualities of drinks and food, wines and meat. And on the fourth page of this chapter he states:

> The guests at the *Symposium* hiccup, speechify or slump about, weighed down by alcohol, Plato has ensured that the banquet never takes place. They speak of love without making love, sing of this or that without actually singing, drink without tasting, speak with the first tongue – but for all the sounds they produce, do we know what wine they drank: from Chios, Corfu or Samos? (Serres 2008: 155)

Serres accuses Plato here of simply missing the point and of avoiding the actual challenge. This challenge would have been to actually think precisely about these major entities of life – for example loving, singing, drinking, eating – *while* practising them; and not while not practising them, but merely resting on some comfy cushions, 'reclining on a divan like a god, its cup always left untouched, a robot with an anaesthetized mouth, its parts of marble or metal, indifferent, empty, punctured, stoppered, absent' (Serres 2008: 225). Serres proves Plato wrong by an argument that is unknown if not completely incomprehensible to the discourse of philosophy – an argument he shares then with Eshun: the argument of situated, experiential and corporeal consistency. An argument focusing on the body of the researcher. This argument might be discarded as a pseudo-rational form of *sophism* – be it an *argumentum ad lapidem, ad lazarum, ad oculos,* or the pragmatic argument; yet, both authors argue that theoretical reflections are only as good as their potential to be upheld in the actual situation of practice they refer to. Indeed, this is a deeper and a more complex, more constrained test of practice, a reality check as mentioned before – that implies foremost a critique of an accelerated and often self-serving discourse in research: an almost autopoietic and non-referential discourse that seems to be more often than not radically void of sensibilities and experiences regarding the aptness in a situation.

The chapter 'Animal Spirits' with its roots in food, in preparing food and wine, in eating and drinking, in getting drunk and falling in love, in confusion and confluence, in the generativity of stirring and kneading, of cooking and fermenting, of mixing and crafting – this whole chapter is an infusion of thinking with the spirits of animals, in the zoological, in the mantic, but also in the distillatory sense. The body of the researcher is revitalized and respirited, respiritualized, replenished with all of these animal forces, drives, sensibilities. Serres's epistemology is a corporeal, a carnal and an experiential philosophy. In the strictest sense, the philosophy of Michel Serres is a *philosophy against philosophy*, a thinking against propositions. It is not yet a non-philosophy as François Laruelle would ten years later set out to propose; but Serres pursues a comparably close goal to reintroduce the resistance of materials, of experience, of work and dirt, of bodies and sensibilities, of entanglement and of mingledness – maybe even of mixillogic – into an area where still the platonic desire for clean and distinct tables of terms and concepts and subsequent clear decisions dominates. Serres brings the body back into research. In all these respects, Michel Serres's *The Cinq Sens* is not just a book written *about* the senses: *The Cinq Sens* is a book written *out of* the senses – as is also *More Brilliant than the Sun*.

Syrrhesis Fiction

Serres's and Eshun's writing clearly bears common traits, partially common goals, and – at least punctually – also common strategies of arriving at their goals. Both approaches band together in a coordinated attack against traditional and, as some would claim, sclerotic methods of researching, thinking and writing on the one side, and to proposing a provokingly new and deviant way of writing about propositional contents on the other side. They agree primarily on three aspects: both focus on visceral and material effects and reactions regarding their research issues, objects or non-objects; both favour an unfolding of experiential practices that are largely different from the related literary or philosophical concepts that were extensively discussed previously; and finally, both argue for an expansion of imagination, obsessions and fictions in their writing as major epistemological techniques. In two other aspects though both authors differ distinctively: in their understanding of technology and the role it should play in a future society and how they assess the role of humanoid experientiality as such in this development. But let us start first with the common traits in their thinking and writing.

Both Serres and Eshun root their argument – regarding the first shared trait – in visceral and material effects and reactions. Both authors claim that traditional forms of literary theory or, respectively, philosophical discussions are in general not sufficient to explicate their individual research issues, objects or non-objects, which means the sonic and sensory experiences in a humanoid alien's life:

> Socrates, Agathon and Alcibiades speak of love without ever making love, or sit down to eat without actually eating or drink without tasting; likewise they enter directly from the porch, over the threshold, into the dining area, without ever visiting the kitchens. Like the Gods, slaves and women stand near the stoves, where transformations occur, while the barbarians talk. (Serres 2008: 165)

On the contrary both authors propose – and this is the second aspect they share – to precisely reflect through such experiential practices in their enquiries. Practices that occur or are performed in the kitchen or the sea ship, in the recording studio or the dance floor, in film (post-)production or in music recording facilities. As surprising as this might sound, both Serres and Eshun are *thinkers of praxis*: they make an effort to include the specific experiences, the intricate and detailed sensory knowledge, the embodied forms of epistemology from their reference fields – for example cooking, oenology, music or film production – into their reflections as method. Their arguments are not guided by ancient axioms, questionable truisms or some outworn literary or philosophical concepts; the arguments by Eshun and Serres rely on actual, intense and often erratic experiences. This connects both of their approaches fundamentally, for instance, also to Henri Lefebvre's demand for a new discipline he proposed – following Gaston Bachelard (Bachelard [1950] 2000; Pelleter 2018: 48–50) – to call *Rhythmanalysis*:

> The rhythmanalyst calls on all his senses .... He thinks with his body, not in the abstract, but in lived temporality .... He garbs himself in the tissue of the lived, of the everyday. (Lefebvre 2013: 31)

Finally – representing the third aspect of their shared goals – both thinkers assign in their writing a central role to individual imaginations, obsessions and fictions as major epistemological techniques. Therefore, both are most definitely not traditional or reductive materialists that would deny the existence or the role of desires, dreams and imaginary worlds; on the contrary, both Eshun and Serres reserve for them a central role, be it as afrofuturist and technoutopian imaginations in the writings of Kodwo Eshun – or as sensory encounters with the artefacts of art history, of biographical narrations, or particular cultural environments and new cultural practices in the works of Serres. Both operate, hence, foremost in the framework of a *new materialism*.

All these three common aspects I would like to subsume in the neologism of a *Syrrhesis Fiction*. This new term combines Eshun's concept of *sonic fiction* with Serres's method of *syrrhesis*: in combination a Syrrhesis Fiction would thus represent a thinking that operates through, in and with practices – a thinking out of practices. Because Serres proposes to replace the scholarly and analytical approach of trying to understand, to scrutinize and to atomize a given phenomenon, with a synthetical (or *syrrhetical)* approach. This approach now constitutes the core of practices such as cooking, making wine, painting, writing poetry or producing music, mixing tracks or crafting a sound installation in a given architectural space. For Serres, to put it bluntly, any analytical knowledge – of foods, of sounds, of spirits, of visuals – needs to be regarded as inferior to the actually practised and skilfully performed craft and art of working with precisely this knowledge – in preparing foods, sounds, spirits, visuals. Analysis succumbs to syrrhesis.

The thinking of both authors is, therefore, not proceeding through analytical tables or decision trees, but through the practice and the reflection on crafts and situations, on the effects of visceral and technical constellations, on the imaginary and the fictions attached to these visceral crafts. Serres and Eshun invent, they both dream of a new specimen of reflection and research, of epistemology and of imagination:

> We dream indistinctly that a word capable of expressing this confluence might be acclimatized into our tongue. We cannot say concade nor syrrhesis. (Serres 2008: 161)

Eshun explores the sonic fiction of uncharted, suggestive crafts of sonic afromodernity and Serres explores the syrrhesis of still widely uncharted experiences and sensibilities. A thinking of *syrrhesis fiction* would then not any longer be restricted to arguments relying on verbal citations or conceptual terms; but this thinking could, ideally, integrate visceral experiences and everyday practices, sensory encounters and sonic affects in its fantastic and imaginary arguments. This common effort by Eshun and Serres makes them pioneers and experimentalists of future epistemologies and, if you will, *xenonthropologies*. As these common traits and shared goals are clearly given, there is though quite a number of aspects in their writings where the common aspects are more complicated to find or to extract, to say the least – left aside all their obviously and massively differing biographies, acadamic upbringings and their underlying professional practices: DJ Kodwo versus Seaman Michel. Both authors differ then primarily in their understanding of technology and the role it should play in a future society. Eshun regards, almost needless to say at this point, technology as a major force of progress in an accelerationist line of arguments, apparently propelled by technoutopian hopes of futurism and scientific optimism:

> Where crits of CyberCult still gather, 99.9% of them will lament the disembodiment of the human by technology. But machines don't distance you from your emotions, in fact quite the opposite. Sound machines make you feel more intensely, along a broader band of emotional spectra than ever before in the twentieth century. (Eshun 1998: -002)

> From now on, Electronic Music becomes a technology-myth discontinuum. Traditional Culture works hard to polarize this discontinuum. Music wilfully collapses it, flagrantly confusing machines with mysticism, systematizing this critical delirium into information mysteries. (Eshun 1998: 161)

In harsh contrast to this, Serres regarded as early as 1985 the development of a large-scale computer society relying on data mining and modelling of a *word* or of *dictionaries* as dangerous and misleading. He argues even for a thoroughly experientialist and deeply sensorial education:

So, do we learn how to die, how to survive alone through
suffering, to sing joyfully when our child recovers from illness,
to prefer peace to war, to build our home over time? Or do we
take our education in the direction of serenity? In dictionaries,
codes, computer memory, logical formulae; or quite simply at the
banquet of life? I don't believe, says the beggarly phantom behind
the machine, that if there is any sense to life, it lies in the word life;
it rather seems to me that it arises in the senses of the living body.
Here, in the sapience cultivated by fine wine, with as few words
as possible; in the sagacity mapped out by scents enhancing our
approach to others; there, through vocalizing, sobbing, and what
our hearing perceives beneath language; through the aromas
that rise up out of indescribable earth and landscapes; from the
beauty of the world that leaves us breathless and speechless; from
dancing, where the body alone dives freely into deaf and mute
senses; from kisses which prevent us from even whispering… from
the banquet we will have to leave. (Serres 2008: 195–196)

More recently though, Serres recognized even in an optimistic
tone the potential liberations and new life forms, new experiences
emerging from our meticulously digitized, mediatized and
networked societies:

Without us even realizing it, a new kind of human being was born
in the brief period of time that separates us from the 1970s. He
or she no longer has the same body or the same life expectancy.
They no longer communicate in the same way; they no longer
perceive the same world; they no longer live in the same Nature or
inhabit the same space. Born via an epidural and a programmed
pregnancy, they no longer fear, with all their palliatives, the same
death. No longer having the same head as their parents, he or she
comprehends differently. (2014: 7)

Both authors finally then also differ in how they assess the role
of humanoid experientiality as such in this development. Serres
indeed puts a lot of conceptual effort and hope into detailing
how humanoid experientiality is effectively superior to humanoid
commodity cultures. He even regards individual experience as
the one major ressource for future developments and for social
progress, whereas Eshun is much more sceptical regarding such

rather anthropocentric views. Though Eshun also articulates his hopes towards social and cultural progress through machines, through technology and their autopoiesis. Both authors concede, dialectically trained, the interdependency of both categories and areas of cultural development, technology's autopoiesis or humanoid experientiality – but their individual and primary focus of reflection and research stays distinctly different. Eshun remains a technoutopian though not denying the gargantuan dystopian double-bind inherent to all technoeschatologies; Serres remains anthropocentric while never ignoring the maybe surprising potential of all the inventions emerging out of new technologies and even commodities. Mixologically, it seems, both can be read as complementary pioneers and inventors, joining forces to generate and to promote syrrhesis fictions – to generate new epistemologies.

Beyond the Idiosyncrasy of Logocentrism

Dancing is an epistemic practice. Walking is an epistemic practice; eating is an epistemic practice. Drinking is an epistemic practice. Smelling is an epistemic practice. Touching is an epistemic practice. Listening is an epistemic practice. All of these practices and many more I did not list here are *research activities*. They form a genuine – I'd even dare to say: a crucial – part of the history of humanities. Without these activities researchers simply would not be doing research at all (cf. Schulze 2016). Epistemologies in general focus on knowledge production, on the established practices of distinguishing between legitimate and illegitimate research methods at a given historical time and a given institutional framework, and on the procedures and dispositives of academic confirmation or rejection of research results. A *sensory epistemology*, therefore, transcends the existing and legitimate institutional and historical framework of academia; a framework that is still focusing its operational modes on sign operations, on definitions of terms, on decision trees and on propositional sentences. The apparatus of logic, of syllogisms, and of calculation still primarily defines what is recognized as a *proper research practice*.

A sensory epistemology – and similarly an artistic, an aesthetic or a visceral one – exceeds these limits of signs and their processing.

Research then can take place in virtually any imaginable situation of everyday experience, of corporeal activities, or of a professional craft. Cooking or dancing are not less epistemologically valid than counting or interpreting. The main element that distinguishes a sensory from a traditional epistemology though is the fact that it does not rely mainly on written, formulaic or calculated accounts, printed or handwritten, of its results along the lines on the surface. These new epistemologies rely on a bricolage, an accumulation, or a meshwork of sensory experiences; on a situated craft – or on material and aesthetic performativity. Now, what might sound like a weird, alien and thoroughly idiosyncratic decision to move away from sign operations to a wild and erratic mixology of sensory experiences is precisely the opposite: it is an effort to leave the insanely idiosyncratic, only historically legitimated decision for an alphanumeric and logocentric epistemology of writing cultures and *Aufschreibesysteme* (Kittler 1985) behind. The goal to incorporate the full and yet inexhausted complexity of experiential constellations and mixtures, of all the divergent idiosyncratic selections into the concept of epistemology, this goal multiplies only the existing *Idiosyncrasy of Logocentrism*: drinking, smelling, touching, listening, loving, breathing can be as epistemologically insightful and idiosyncratic as writing, calculating, interpreting, drawing. These new efforts of epistemologies therefore resist and reject an unnecessarily limited idiosyncrasy of logocentric epistemologies. They break out of the black prison of signs and characters into the vast and rich potential oft yet unexplored and unassessed new, sensory and multiple epistemologies. *Multiplestomologies*. Not that they are idiosyncratic and stubborn and weird: they are polycentric and dynamic, versatile, agile and transformative. The existing and hegemonic epistemology to the contrary represents a by now unnecessary and compulsive reduction and limitation of epistemic potential.

However, to explore this immense epistemic potential it needs a *generative* epistemology that does not confine itself to just confirming the existing epistemic practices. Serres and Eshun both take on this endeavour. By doing this they again make the by now well-established characteristics of a sonic fiction prolific: mythscience, mixillogics, mutantextures. In generative epistemologies the *mythscience* according to Eshun becomes a transformative force; this force then exceeds the mere realm of one research object – such as music, composition, sound art or sound performances.

The transformative force of mythscience is here invading, virally infecting, and changing the shape, the practices and the goals of research itself. One might feel the ground slipping away underneath one's feet. This new epistemological mythscience focuses on aspects usually repressed, neglected or simply ignored – such as 'qualities of sound and tendencies in movement and perception' or, in general, 'what the sonic body already knows' (Jasen 2016: 14):

> This is where sonic fiction overlaps with the other element of mythscience – what Deleuze and Guattari term *nomad science* (or sometimes *minor science*). Nomad science (vernacular and 'problematic') is set against Royal Science (official and 'theorematic'), although the two are essentially linked, diverging in tendency, but always feeding each other. While institutionalized science employs transcendent Method to extract generalizable laws from nature, a more ambulant science works intuitively and contingently, pursuing variation and anomaly, inhabiting materiality and following its singular flows. (Jasen 2016: 14–15)

Intuitively and contingently, pursuing variation and anomaly, inhabiting materiality and following its singular flows: this is how sonic fiction and also how Serres's syrrhesis operate. Generative epistemologies include as syrrhesis fictions therefore such mythsciences. They follow:

> Sonic materiality in signal, flesh, machine and space, intuiting what it can do, experimenting, pursuing anomaly and tweaking things towards qualitative change. (Jasen 2016: 15)

This *experimenting*, this *tweaking things towards qualitative change* – as Paul Jasen describes it – extends epistemologically to the second characteristics of sonic fiction: *mixillogics* embody the syrrhesis in its confluence – in Serres' terms – its open-ended and searching recombination, the excited trial and error, the freaked out and joyful mixing in of ever more different and new and unknown substances and qualities and practices into the process. Whereas a mythscience represents the divergent and non-normative, the exceptional character of this epistemology, then the mixillogic represents the effort to craft the most surprising and unconventional and experimental constituents of praxis into research. This is what

Serres looked for – as quoted in the previous section – when he somewhat helplessly mourned that we 'cannot say concade nor syrrhesis' (Serres 2008: 161). With Eshun now, he could indeed say: *mixillogic* or *mixillogique*. A word that is defined by Eshun to describe exactly this wild and erratic mixture that might seem irrelevant or weird from the outside – but actually serves a clear purpose in the pragmatic sequence of trials and experiments, of all the activities that are intended to generate a new, maybe again surprising, but surely generative *mutantexture*.

When Eshun writes in *More Brilliant than the Sun* of such mutated textures of sound and sensory events, he is indeed unfolding precisely the 'phenomenological possibilism' (Voegelin 2014: 48). Salomé Voegelin discusses that by sensory or material explorations new sonic or epistemic possible worlds are being generated, materialized, triggered. The epistemic phenomenology of *mutantextures* is best exemplified by all the journeys into various sensory epistemologies performed by Michel Serres in his writings. One of the most radical endeavours to open up a possible epistemology in deviation from traditional European philosophy can therefore be found in his aforementioned rewriting of Plato's Symposion by including all the corporeal lushness and delicacies, all the indulgently detailed descriptions of furniture and clothings, bodily reactions and carnal desires, of kissing and hornyness, of drunken ramblings and intoxicated stupid dancing. All the details of the material and crude reality of a binge-drinking feast with chatty patriarchs that were so neatly concealed and paraphrased by Plato's idealizations get here to be exposed and unfolded:

> Empiricism takes refuge in the kitchen alongside the kitchen boys smeared with sauce, and the maids, saucy brunettes in white aprons. Quite well-behaved, even simple-minded, it listens to the speeches after the wine, takes fright at the jovial, booming actors, hams, prostitutes, imperious and decorated as they are. It is frightened of philosophy, science and laws, preferring to withdraw. To leave the table before the end. (Serres 2008: 230)

> It has indeed taken the whole history of philosophy, which from its very beginnings had nonetheless intuited mixture and chaos, to rediscover in a glass or a vessel, in a simple, naive, almost childlike way, what was already happening in the kitchen while the guests drank and spoke of love, and what vignerons

have been doing in an insanely complex manner since the very beginnings of our traditions. (Serres 2008: 168)

In Serres's rewriting of the symposion the materialist and visceral details finally come about. This new and actual symposion includes the mythsciences of drinking and cooking and tasting. Syrrhesis and mixillogics are finally allowed to happen through intoxication and conversation, through carnal pleasure and erotic sensibilities, humanoid orifices of digestion, of desire, of speaking in elevated, delirious tongues. These are the syrrhetic practices that constitute, supposedly, the actual symposion and its factual mutantexture. Any logocentric reductionism seems now so far away from here. Who would, under these sensual circumstances, wish to reinstate the scaremongering and regulatory practices of excluding and of sanctioning the corporeal, sensorial, the situated and idiosyncratic aspects of this issue or that argument. All the *Uses of the Erotic* are present here:

> To refuse to be conscious of what we are feeling at any time, however comfortable that might seem, is to deny a large part of the experience, and to allow ourselves to be reduced to the pornographic, the abused, and the absurd. (Lorde 1984: 59)

As unfolded in Audre Lorde's crucial essay – published around the same year as Michel Serres's – only with such a dismantling of this corporeal, sensorial and erotic reduction can one move towards an inclusion of all these materialist and visceral, those highly malleable and relational qualities in living, sensing, in crafting, and also in researching. When all of these sensory qualities are mixed into research in a process of syrrhesis with a sensibility that one might also rightfully call then *erotic*, substantially consituting the mythscience, they generate altogether new mutantextures:

> To intensify sonic experience, to rhythmically vary it, producing surprises and actualizing things previously only imagined. Sonic fictions theorize becomings and conceptualize affects; they attempt to find language for the mystifying feeling of affect's escape – the sense that one is caught up in more than meets the ear, and that reality doesn't quite add up. Their companion is an inductive science, comprising technical practices and techniques of *affect engineering*, designed to draw people out of themselves and into an unfamiliar relation. (Jasen 2016: 15)

Lorde and Serres and Eshun meet in this very quest for a generatively sensory, a corporeally epistemological practice.

Multiplying Epistemologies

Positioned in sound, in its visceral and material impact, one moves away from a traditional and distant epistemology. An epistemology that imagines some anonymous, objective, omni-erudite and all-knowing researcher as its steering entity – always male and athletic and always in charge, 'white, thin, male, young, heterosexual, christian, and financially secure' (Lorde 1984: 116). A strange and actually inexistant 'mythical norm' (Lorde 1984: 116). From this epistemic idealism one moves carefully, daringly and curiously into epistemic materialism and realism. A sensory and sonic materialism that materializes actual and existing sensibilities and subjectivities of experience – in all their glorious erratic richness:

> Sonic materialism is not objective, but produces subjective objectivities, the materialities of private life-worlds, from which we negotiate contingently the material form of the world. (Voegelin 2014: 100)

Such an approach allows through its intimate interweaving with specific materialities, its mythscience, mixillogics and mutantextures to generate time and again new epistemologies. Syrrhesis fictions bloom and bloom. This ground in syrrhetic mixillogics liquefies all epistemic desires and experiences. It multiplies the potential epistemologies. They do never really stop and solidify, they are never actually finished, they continue to generate new specimens, hybrids and variants and versions of epistemologies. Each epistemology, emerging out of a sensory encounter, generates anew a quite different world. A new and diffracting world that helps to answer one recurring question:

> *In what ways could we imagine a world different from the one in which we currently live?* (Gunkell, Hameed & O'Sullivan 2017: jacket copy)

The open and necessarily so generative epistemological practice and thinking of Michel Serres leads to this multiplicity of epistemic specimens – and also to an as large multiplicity of forms to articulate, to demonstrate, to teach or to present. However, this approach requires a lighthearted crossing between the differing and heavily guarded borderlines of publishing and of researching – be it in the arts and in fiction or be it as a state-employed professor. Precisely this strategy Dietmar Dath most recently recommended for multiplying epistemologies and mutantextures:

> Topographies of cognition such as the scientific, the philosophical and the aesthetic can be expanded by reconstructing each of them in one of the others. If you do this in writing, you have to write treatises as well as stories, poems as well as manifestos, analyses as well as speculations – namely poems about analyses, speculations about stories and so on. (Dath & Greffrath 2018: 33;[1] translated by Holger Schulze)

Mixillogics multiply not only into an endless variety of mutantextures but also into the possible worlds they imply. These possible worlds of all potential and particular futurisms are inherent to science fiction as they are to sonic fiction or to syrrhesis fiction. It goes without saying that possible worlds of post-binary gender and their myriads of intricate sensibilities are equally included here. Multiplestomologies grow and hybridize in all directions, on all layers, in all dimensions and dynamics imaginable:

> 'She' is the pronoun for all sentient individuals of whatever species who have achieved the legal status of 'woman'. The ancient, dimorphic form 'he', once used exclusively for the genderal indication of males (cf. the archaic term *man*, pl. *men*), for more than a hundred-twenty years now, has been reserved for the general sexual object of 'she', during the period of excitation, regardless of the gender of the woman speaking or the gender of the woman referred to. (Delany 1984: 78)

5

Acid Communism
A Haunted Utopia of Sound

Hauntology is the proper temporal mode for a history made up of gaps, erased names and sudden abductions. (Fisher 2013: 52)

Throughout the 20th century, music culture was a probe that played a major role in preparing the population to enjoy a future that was no longer white, male or heterosexual, a future in which the relinquishing of identities that were in any case poor fictions would be a blessed relief. (Fisher 2014b: 28)

Behind a grey veil of distant listening, somewhere across this meadow, across this pond on a countryside we can hear the bleeping sounds of analogue synthesizers, muffled drum loops, fading lullabies, somewhere, somehow. Is this the past reimagined now? Or is this the future nostalgically reminiscing a past when artists still imagined a better future, a progressive future? A future that never was? The realm of fiction is a realm of multiple time regimes, of overlays and underlays, of double and triple and quadruple exposure of experiences and situations and moments – all together in one place, one second, one instant. In fiction the common and orderly sequence of time which you and I have learned to follow, time and time again in our childhood and adolescence, to obey, to conform, partially, precisely this sequence is broken up again and opened up again and questioned later in life, questioned in fiction. There simply is no given sequence of time in experience. Memories haunt us, fears terrify us, self-consciousness can paralyse us, drug-infused furores of cocky self-aggrandisement and self-elevation can superinflate and hyperaccelerate us. One can hear these intersecting time regimes in the music of Boards of Canada, one classic example for hauntology in music: *Music Has the Right to Children* from 1998 (cf. Reynolds 2011: 330–335). The past is never, truly *never* over – it never actually ended. The future is always already anticipated in a multitude of utopias and dystopias – it always has been with you and me. This presence right now is not always as vitalistically experienced as the ecstatic propagandists of *Living In The Moment*™ paint this presence in their weekend seminar keynotes and globally streamed TED-talks. There are as many experiences of presence as humanoid aliens experiencing a presence, at minimum. In 2012, Kodwo Eshun writes about Dan Graham's video essay *Rock My Religion* (1984) in its mixillogics of historical periods and social formations regarding:

> An idea of America – a construction that starts with the religious communities that left the England of the Industiral revolution (and even earlier) for the New World, and that finds a culmination of sorts in the social formations that emerged after World War II, shaped by new urban structures, mass cultural production and unprecendented forms of consumerism. (Eshun 2012: 3)

It is precisely this layered, overshadowed and deeply mingled and entangled quality of experience and of fiction, of the empiricist epoché of this very moment, right here, right now, incorporating all the manifold imaginations and fictions and artistic constructions in this one moment that connects Eshun's writings to the writings of Mark Fisher. This underlying connection was unfolded in Fisher's trilogy consisting of *Capitalist Realism* (2009), *Ghosts of My Life* (2014) and *The Weird and The Eerie* (2017). Therein Fisher explores the interdependencies between capitalist life conditions, experiential sensibilities, their articulations, and the actual effects this explosive mixture has on everyday life and on contemporary culture. His journeys though bring him ever closer and closer to the structurally dystopian, the thoroughly technologically infested writing and thinking of Kodwo Eshun. As an existential and painful consequence, one year after Fisher's suicide on 13 January 2017, Eshun was the first speaker invited to hold the Mark Fisher Memorial Lecture at Goldsmiths, on 19 January 2018. Herein

Eshun characterized Fisher's work to a large degree by its sustained and ongoing effect on his colleagues, friends, his readers and his disciples:

> What matters for those of us alive – now, on January 19th, 2018 – is to work out the ways and the means and the methods for continue to work in and with, and away from, and by way of Mark's writing and his thinking. A thinking which is inseparable from his enthusiasms, from his impassioned thought, from the polemical determination he brought with him. (Eshun 2018a: 6:09–6:55)

Eshun stresses the qualities of the energy of engagement, of forming movements, of inspiring interpretative communities, impactful discourses and caring for a progressive development of society and culture in his admirable eloge to Mark Fisher. They represent some of the values he also holds dear; even if he might have realized at some point that Fisher as a scholar and an academic working in a major institution, unlike himself, is much more capable of enacting them on a social and institutional level. His own work focuses to the contrary more on to the artistic and conceptual explorations of the conditions of possibility for such progressive developments held dear. At the end of their shared vision stands a term that probably both authors would claim as a rather attractive utopia – a realistic utopia: *ACID COMMUNISM*. This then became also the title of an introduction to a collection of essays by Mark Fisher on which he worked in the last months of his life (Fisher 2018: 753–772): a goal to evade from contemporary restraints and pains and restrictions in society, economy, politics and culture. A goal that qualifies easily as a major motivation for both authors. A goal with which both authors assume they could find refuge from the weird and the eerie moments that engulf us, representing in a haunting way also the void, the lack of this very goal. The sounds and the sensory events both authors describe as eerie and as weird in their writings, they might therefore grant a glimpse into this as yet not attained goal; a time when most contemporary experiences could be described as examples of a 'boring dystopia' as Fisher called it (Kiberd 2015). In Macon Holts inspiring study he outlines the area of sonic fiction regarding precisely these interferences between *Popular Music and Hip Ennui*. The element of science fiction in sonic fiction though is for Holt more

represented by the dystopian approach, for instance, of Philip K.
Dick, J.G. Ballard or David Foster Wallace:

> A science fiction set in a world long after Baudrillard's semiotic
> apocalypse (Fisher 2000) in which signs are the only thing that
> can hold value any longer. To use another term of Fisher's, we
> need a science fiction for our 'boring dystopia' of capitalist
> realism. And this is how I read Wallace's *Infinite Jest*, a novel
> about the desperate search for meaning in a world where even
> time as such has become a commodity. . . . If we can hold to
> this apparent but not actual conflict, which is derived from
> the condition of capitalist realism's boring dystopia, as we
> approach the Sonic Fiction of contemporary pop music, we
> may be able to move past the drive to neoliberal conformity
> that prevents the politics in this music from being heard.
> (Holt 2020: 108)

This is the situation of existential loss, of lack, of an all-
encompassing void that is the actual starting point for Fisher
and in part also for Eshun: 'a lost utopianism: the post-welfare-
state era of benevolent state planning and social engineering'
(Reynolds 2011: 330). However, this painful connector has
apparently served – and it still does – as a surprisingly strong
attractor for a large number of activities by so-called *interpretive
communities* Eshun finds attached to Fisher's ideas and thoughts.
The pain here is facilitated as a common ground and shared
experience that Alain Badiou articulated in the aftermath of
Donald Trump's election as the president of the United States as an
existential and political impasse:

> We have no government in the world which is saying something
> else. And why? Why, finally, if we examine the position of the
> 'socialist' French government, of the *dictature* [dictatorship] of
> the Communist Party in China, or the government of United
> States, or the government of Japan, of India, everybody says the
> same thing — that globalized capitalism is the unique way for
> the existence of human beings. (Badiou 2016)

The scattered communities that do still feel some nagging and
hurting doubt concerning this seemingly one and only truth, they

find themselves haunted by sounds. Sounds that represent a lost, a sclerotic, a buried utopia.

> Sound set the terms for looking not in order to underline psychological territory nor to act as musical character but to shape the contours of [some] terra incognita. (Eshun & Sagar 2007: 95)

Anticipation and Compulsion

In the writings by Fisher the lost utopia of a desired future beyond one's current pains and losses, desires and longings, this imaginary world and future world is recurrently mourned and encircled. Fisher takes this as a starting point for a critique of contemporary impasses in society, politics, economy and culture – and, in a second step, also as a starting stroke to draw a sketch of possible worlds we might be moving towards, if only tentatively:

> Philip K. Dick could have predicted the banal ubiquity of corporate communication today, its penetration into practically all areas of consciousness and everyday life. (Fisher 2018: 756)

In his diagnosis Fisher is truly a disciple of the CCRU and its critique of the effects and potentials of digital culture. One can even hear the distinct resonances from Nick Land's dystopian (and in the end neoreactionary) interpretations of society's and culture's recent developments. Fisher writes in his article 'What Is Hauntology?':

> What haunts the digital cul-de-sacs of the twenty-first century is not so much the past as all the lost futures that the twentieth century taught us to anticipate .... More broadly, and more troublingly, the disappearance of the future meant the deterioration of a whole mode of social imagination: the capacity to conceive of a world radically different from the one in which we currently live. It meant the acceptance of a situation in which culture would continue without really changing, and where politics was reduced to the administration of an already established (capitalist) system. In other words, we were in the 'end of history' described by Francis Fukuyama. (2012: 16)

This diagnosis of an existential cul-de-sac describes precisely the *hip ennui* Macon Holt speaks of. This sentiment of being imprisoned in this present and of having lost all utopian and as such empowering visions of refuge, of resistance, of subversion, this precise sentiment represents a constantly bitter feeling of defeat: a defeat in which the anticipated glorious futures of the past, depicted and sonified, imagined and sculpted in a long, seemingly endless series of futurist artworks, compositions, movies, of novels and designs, a defeat in which this anticipation as well is lost, is ridiculed and discarded onto the ash heap of history. Not only your personal life is destroyed and horrible, even all your hopes and dreams and potential, imaginary worlds have become meaningless and hollow. Life right now though can be lived and experienced, one can still act and perform, breathe, eat and speak, consume and metabolize. But there's no meaning in there any longer. Attached to all the things, people, institutions, processes and projects, there is now only the capitalist real: capital accumulation and operations for profit. A deserted and void cosmos. This is the world of melancholia. A melancholia that arises, such as Lars von Trier's (2011) apocalyptic exoplanet, from the overwhelming consciousness of a transformed existence. The melancholia of the capitalist real haunts this very planet, compulsively. In *Ghosts of My Life* Fisher mourns the missing solidarity amidst all of our *marvels of communicative technology*:

> One way of thinking about hauntology is that its lost futures do not force such false choices; instead, what haunts is the spectre of a world in which all the marvels of communicative technology could be combined with a sense of solidarity much stronger than anything social democracy could muster. (2014a: 26)

This anticipation of a potentially better world and at the same time the compulsion to remember this while fully aware of its actual impossibility, this emotional and sensorial paradox, this oxymoron is, what Fisher coined as the *hauntology* of our times. It represents a vertigo, an almost unbearable feeling of being on the one side compulsively haunted by a utopia lost and on the other side still anticipating exactly this desired future of a lost utopia: a doubled and intertwined sensation of a loss embedded in hope. Now, precisely this complex affect can be brought in connection to communist theories of the ghostlike virtuality of a better world, following Fisher:

> The first refers to that which is (in actuality is) *no longer*, but which *remains* effective as a virtuality (the traumatic 'compulsion to repeat', a fatal pattern). The second sense of hauntology refers to that which (in actuality) has *not yet* happened, but which is *already* effective in the virtual (an attractor, an anticipation shaping current behaviour). (2014a: 19)

Anticipation and compulsion oscillate at the core of hauntology: *compulsion to repeat* and *an attractor shaping current behaviour*. This affect is neither clearly cut nor strong and distinct; its main characteristic is indeed its murkiness, its entanglement, its hopelessness, also its meagre and yet inescapable double-bind. An affect sounding as far and away, awash and veiled, through rain and vinyl's static, through pitched ghostlike voices and basslines of imminent doom, as in Burial's debut album *Untrue*, from 2006. Compulsively returning to the same old memories of gloom, anticipating in its sound production a future that never was:

> Tell me I belong, tell me I belong, tell me I belong
> Holding you
> Couldn't be alone, couldn't be alone, couldn't be alone
> (Burial 2006: track 2)

Haunted by a lost love, by haunting sounds. Being haunted by such strongly anticipated and compulsively repeated hopes for a better future is a grey and blurred and gooey sensation. A sensation of pervasive melancholy:

> The kind of melancholia I'm talking about...consists, that is to say, in a refusal to adjust to what current conditions call 'reality' – even if the cost of that refusal is that you feel like an outcast in your own time. (Fisher 2014a: 24)

Hauntology bears from one perspective, assumed at the beginning of this section, the quality of an anticipated hope lost and the compulsion to remember it. From another perspective though – tangible maybe just now – it is primarily characterized by the compulsive remembrance of this very utopia lost, with only a slight chance to anticipate it through its melancholic presence. The paradox of a *loss embedded in hope* turned into an almost motivating combination of *commemoration as conjuring*. Could this loss be actually generative in the long run? Could it provide an *implex*? Fisher might have implied this when he proposed such hauntological sensations as fundamental for our times, especially the 2000s and 2010s. Starting from this paradoxical experience he then connects our period of radical non- or even anti-communist lifestyles provocatively with the spectre that haunted an earlier historical period in which it later materialized as communist and socialist parties. The ghostlike virtuality of a lost utopia turned, hence, into one of the most famous appearances of a spectre in recent cultural history:

> The 'spectre of communism' that Marx and Engels had warned of in the first lines of *The Communist Manifesto* was just this kind of ghost: a virtuality whose threatened coming was already playing a part in undermining the present state of things (Fisher 2014: 19).

It is this precise void, this painful lack of a stolen and vaporized utopia that conjures such ghosts. These ghosts though do not remain immaterial and ephemeral, fugitive and intangible. First they populate our lives and sensibilities, our thinking and sensing; then they might more and more materialize and take effect in selected and lasting realities. And maybe, just maybe, in a retrospective interpretation the anticipation and the compulsion of our haunted lives might not have been so futile? Maybe our behaviours, our desires, our mournings and depressions have the chance to become generative and transformative – at least at one point in the near future? Even if this might seem hopeless and empty and next to impossible at the very moment, right here, right now, then indeed this conjuring of ghosts might just provide a first, a minuscule step into a then possible and a more hopeful future world. A world that might even be able to prevent a terminal, planetary climate catastrophe? But how do the ghosts of our times actually achieve this?

Ghosts of Our Times

The ghosts of today, the ghosts or *spectres* that populate Fisher's writings were partly born in the nineteenth century, in the *Manifesto*

of the Communist Party – 'Ein Gespenst geht um in Europa —
das Gespenst des Kommunismus' (Marx and Engels 1848b: 4) –
and in the late twentieth century, in Jacques Derrida's *Specters of
Marx: The State of the Debt, the Work of Mourning and the New
International* (1994). If one were not familiar with these spectral
ancestors they might also be understood to represent a return to
premodern and pagan belief systems. The dead haunted the living
for centuries and millennia – and just when modernist non-believers
assumed they had vanished now for good, they returned even more
gloriously around 1900. These ghosts on celluloid, writing on
typewriters and appearing on early photographs were a side effect
of the new and contemporary media technologies and their often
ghostlike appearance, their 'sufficiently advanced technology' that
seemed so 'indistinguishable from magic' as science fiction writer
Arthur C. Clarke noted in his famous essay 'Hazards of Prophecy'
(1962: 30). Ghosts emerge, apparently, out of political desires and
out of ruptures in media history. Ghosts appear when our lives (or
some aspects in them) are *neither present nor absent, neither dead
nor alive*:

> Hauntology supplants its near-homonym ontology, replacing the
> priority of being and presence with the figure of the ghost as
> that which is neither present nor absent, neither dead nor alive.
> (Davis 2005: 373)

The magic and the ghosts in nineteenth-century media technology
resulted on the one hand from the first apparatuses trying to pin
down and to document the very existence and reality of virtually all
entities and their agency in images, sounds, words and movements;
on the other hand it was at precisely this historical moment that
the transition from bourgeois societies of manufacturers and first
industries gained traction and boosted the economic growth,
international connectedness, global traffic and intercontinental
communication systems. This massive acceleration fostered
dialectically at the same time an urge to find comfort and refuge in
ancient and almost abolished habits, belief-systems and unscientific
'*Weltanschauungen*'. In the course of the twentieth century this
haunting threat then materialized in historiography and most
explicitly in black vernacular music as Klaus Theweleit recognizes:

> In historiography, truth and affect-loaded fiction are hard
> (and sometimes impossible) to distinguish... GHOSTS: – 30

years ago that was a piece from the tenor saxophone of Albert
Ayler...highly real...today it is something Michael Jackson
dances on...they've come a long long way.(1998: 7;[1] translated
by Holger Schulze)

The ghosts and spectres of Fisher or Eshun, of Derrida and also
of Klaus Theweleit or Albert Ayler, share an existence as *affect-
loaded fictions*, they represent and they embody a collective and
political *articulation of need*. But what is the more specific kind
of need and fiction that these authors articulate – aside from
a mere pun-like reference to the *Manifesto of the Communist
Party*:

> A *spectre* is haunting Europe – the *spectre* of communism. All
> the powers of old Europe have entered into a holy alliance to
> exorcise this *spectre*: Pope and Tsar, Metternich and Guizot,
> French Radicals and German police-spies. Where is the party
> in opposition that has not been decried as communistic by its
> opponents in power? Where is the opposition that has not hurled
> back the branding reproach of communism, against the more
> advanced opposition parties, as well as against its reactionary
> adversaries? Two things result from this fact: I. Communism
> is already acknowledged by all European powers to be itself
> a power. II. It is high time that Communists should openly, in
> the face of the whole world, publish their views, their aims,
> their tendencies, and meet this nursery tale of the *Spectre* of
> Communism with a manifesto of the party itself. To this end,
> Communists of various nationalities have assembled in London
> and sketched the following manifesto, to be published in the
> English, French, German, Italian, Flemish and Danish languages.
> (Marx and Engels 1848a: 4; emphasis added)

In this introductory passage of the *Manifesto* the reference to a
spectre is an ambivalent figure that has two effects on the text it
initiates: with the figure of the spectre the urgency, also the danger
if not a violent and lethal threat to all haunted by it ('Pope and
Tsar, Metternich and Guizot, French Radicals and German police-
spies') is unmistakenly stressed on first reading, without any further
ado. The use of this figure says to the reader: *Communism Ain't
Nothing To Fuck With*. The demands and the sheer existence of

communist movements, worker unions, parties and all forms of associations and syndications, all of this is stated with a pressing, an unrepentant strength. You cannot chain ghosts, you cannot put them into jail, you cannot torture them, you cannot possibly put them on trial: these spectres – of communism – will now stay with Europe and with the world, for evermore. They do not intend to leave very soon. Their spell is now upon you.

This figure of the ghost though also articulates the precarious and contested nature of this entity: *does all of this really exist? Will communism ever have an actual effect on our societies – be it in the near or far future? Isn't it just some lunatic fantasy, incited by technological inventions and a stresssful worklife? You really think communism isn't more of a delusional disease – but an actual political position?* This second effect of the ghost in the *Manifesto* stresses therefore its ambivalent position being *neither present nor absent, neither dead nor alive*: in 1848 communism was not yet as institutionalized as in 1918, 1948 or 1988 – it was still more a spectre haunting premodern aristocracies and modern industrialized communities. It was a spell cast upon modern societies by Karl Marx and Friedrich Engels. This spell resurfaces then again and again in the subsequent history of political theory as well as in epistemologies and fundamental reflections on the condition of societies, cultures, politics and economies to come and present. For instance when Jacques Derrida, in the wake of the fall of the Berlin Wall in 1989 and the implosion of all nations governed under a state socialist model, he returns in *Specters of Marx* to this very ghost of 1848. Hereby, Derrida stresses in a situation when their existence is again contested and precarious the very urgency of these spectres and the threat they represent. Derrida demands:

> Instead of singing the advent of the ideal of liberal democracy and of the capitalist market in the euphoria of the end of history, instead of celebrating the 'end of ideologies' and the end of the great emancipatory discourses, let us never neglect this obvious macroscopic fact, made up of innumerable singular sites of suffering: no degree of progress allows one to ignore that never before, in absolute figures, have so many men, women and children been subjugated, starved or exterminated on the earth. (1994: 85)

Whereas some readers of this passage now might be inclined to find counter-arguments, differing statistics and other historical accounts, Derrida genuinely celebrates the return of the spectre of communism precisely in the moment of its too obvious downfall. Derrida seizes this moment of a dénouement, of a most radical and humiliating exposure of all the protagonists of state socialism in an appropriately dialectical manner; he seizes it to return without hesitation from these experiments, failed in many and even paradoxically and futile violent aspects, to the ghost of communism, its inspiration and motivation and painful drive and desire. With this move he acknowledges the downfall – and he prepares at the same time for a potential future renaissance of this communist reality. The state socialist constructs might be vaporized for now – but precisely this frees the communist idea to again and even more powerful than before haunt all capitalists who rejoice and mock and exploit and harass and destroy this planet and its populations, humanoids, animals, ecosystems, now with even more self-indulgent and disrespectful, with arrogant and sadistic fervour. The quicker you kill its strange and hybrid avatars, the faster this spectre of communism will again haunt not only Europe, but even more so the Americas, the Africas, Asia, Oceania: *Another World is Possible*.

Theories That Are Embodied

Lost utopias and an inescapable imprisonment into this very present society and its structures, they recur compulsively in Fisher's writings about vernacular culture. It is at the same time an excessively hopeful and joyful writing as it is a depressive and hopeless one, a true hauntology of the then contemporary British society. On the occasion of Brian Eno's *On Land* (1982) in relation to Federico Fellini's *Amarcord* (1973), Fisher writes in *The Weird and the Eerie*, his last book, published in the month of his suicide:

The shift into sound opens up the eerie. There is an intrinsically eerie dimension to acousmatic sound – sound that is detached from a visible source – and one of the most unsettling tracks on *On Land* is 'Shadow', which features a quietly distressing whimper that could be a human voice, an animal sobbing, or an aural hallucination produced by the movement of the wind...an outside that – pulsing beyond the confines of the mundane – is achingly alluring even as it is discomfortingly alien. (2017: 81)

Here, Fisher indeed actually writes a sonic fiction about *an outside that is achingly alluring even as it is discomfortingly alien*. He reflects theoretically on sounds and music and films and culture in a way that does not at all hide his very own and very personal *Bedürfnislage* (Koppe), his state of needs – *pulsing beyond the confines of the mundane*. In this very sense his writing is thoroughly structured by hauntology: the author articulates, constantly, how he is haunted by past memories, hopes and sensory experiences – as he desires precisely this lost reality, still present though in this memory. This lingering presence of a sensation long passed lies also at the core of the concept of *hyperstition* – a concept so commonly used by Nick Land and more recently by Steve Goodman that some might even consider it outworn. In an interview request to define hyperstition Nick Land replied in 2009 with this explication:

Hyperstition is a positive feedback circuit including culture as a component. It can be defined as the experimental (techno-)science of self-fulfilling prophecies. Superstitions are merely false beliefs, but hyperstitions – by their very existence as ideas – function causally to bring about their own reality. Capitalist economics is extremely sensitive to hyperstition, where confidence acts as an effective tonic, and inversely. The (fictional) idea of Cyberspace contributed to the influx of investment that rapidly converted it into a technosocial reality. (Carstens & Land 2009)

A hyperstition, hence, transcends the familiar superstition us normies, us generic consumer citizens might believe in; hyperstition refers to an insanely accelerated and therefore massively more prolific version of a superstition. Following Land a hyperstition can *bring about their own reality* and it belongs therefore to *the experimental (techno-)science of self-fulfilling prophecies*. As a consequence, any cultural concept that obtains reality mainly by its mere discourse impact can be called a hyperstition: the more people believe in such a concept, discuss it, doubt it and struggle to define, to understand, to analyse or to grasp it, the more it becomes a new yet efficacious and material reality. By introducing hyperstition into the academic discourse and by applying and referencing it continually, both authors, Land and Fisher articulate as well a performative and malevolent critique of the academic discourse itself. This includes the well-known discourse phenomenon of a certain self-amplification of particular statements by particular actors in the field, by discoursive and *interpretive communities* (Eshun 2018a), as well as the inherent dynamics of contested discourses. It includes also the wider cultural processes extending to mass media effects, journalism's impact and the explosive nature of social media artefacts and fakes in the early twenty-first century.

Some readers might now be inclined to lefthandedly deconstruct or reject this idea as being not very useful for academic research or critical theories. However, the long tradition of reflections on the constructivist character of arguments is hard to deny – ranging in the modern high times of theory from Charles Sanders Peirce's semiotic constructivist approach to reaching a consensus on the one hand to the resignative optimism in Friedrich Nietzsche's writing regarding the role of metaphors that are attracting, guiding, leading and even building one's supposedly very own thinking on the other hand. The most ambitious theorists, thinkers and researchers had to gradually accept during the implementation of a modern research culture and its ongoing discourse in the last three centuries the manifold teachings of critical and self-reflective thinking: what one might have been tempted to call *The Truth*, or an *Insight,* or even *Knowledge* emerges to a larger extent from existing metaphors, from individual predispositions, inclinations, from structural biases, situated diffractions and one's very own desire to uncover hidden and exciting connections and interpretations. Academia is guided by hyperstitions; and so are most areas of the sciences and of all sorts of professions that rely either on public or on secluded discourses.

Here, yet another ramification of the concept of theory-fiction, and therefore also of sonic fiction, can be found: theoretical reflections are not at all restricted or even just constrained to propositions and their logical or convincing chaining and forking. Moreover, this sheer multiplicity of articulations, of formats and representations of theory is not arbitrary. Eshun exemplifies in his Memorial Lecture for Mark Fisher that this multiplicity is actually a main motivation and format of Fisher's 'writings, his criticism, his blogposts, his mixes, his essays, his audio essays, his interviews' (Eshun 2018a: 7:25–7:40). These ever more multiplying options for *theory writing* represent its prolific character, its attachment to phenomena, sensations, experiences in everyday life. Sensory epistemologies generate ever more new sensory formats of academic articulation. The rapidly transforming impact of hyperstitions structuring, deforming and provoking reflections, is articulated in this multiplication of formats.

This is what theory-fictions and sonic or other sensory fictions can achieve, methodologically: they are methods that explore and test culturally influential or marginal concepts by embodying, by enacting, and by even overperforming them. In this respect, both the writings by Mark Fisher and Kodwo Eshun, make an effort through their multiplicity of formats, of approaches and stylistic mutations to take certain hyperstitions of present times, to dive into them, to let them flourish and bloom and expand in all their intricate details and ramifications into all sorts of social and cultural life. However, unlike Nick Land and others, they let these explicative and suggestive narrations also implode now and then by their sheer inflation and expansion, so they can be analysed in their structural weaknesses, their inner contradictions and their inherently tangible false beliefs. Such implosions one would only observe in Land's writings against the author's intention, for example in the crude and stylistically careless patchwork of Land's notorious essay on the *Dark Enlightenment* (Land 2013). Hyperstition for Land is a tool to dominate a discourse at will, but for Fisher and Eshun it is more a descriptive concept to understand all discourse effects driven by the magma of affects and sensibilities that go beyond the supposedly reasonable.

Still, what also Land's writing profits from is this quality of hyperstition that Paul C. Jasen describes as the 'becoming-actual of fictional quantities' (2016: 14). It is a kind of *science theory fiction*: the way a science fiction author, for instance J.G. Ballard, would write and invent and imagine future and yet inexisting forms of theory, maybe in the way Stanisław Lem wrote his imaginary book reviews (Lem [1973] 1985, [1979] 1999) or how Douglas Adams sketched out imaginary scientific theories of the universe and its future encyclopaediae (Adams 1979 amongst others). Fiction therefore contributes a possibly endless multiplication of not only possible worlds but also of possible epistemologies, research traditions and presentation formats. There are as a matter of fact many more research communities and humanoid aliens gathering around cryptic or apocryph research traditions than is often assumed. In his Memorial Lecture for Fisher, Eshun listed then twenty-five interpretative communities that could trace their existence back to Mark Fisher's activity as a midwife for new *forms of life*, new *aesthetico-political positions*, in brief, 'Theories that are embodied' (Eshun 2018a: 23:27). Eshun's list begins with the,

> Cybergoths, that move through the calendrical systems of templexity. The cyber-feminists, that situate themselves in the time-streams of patriarchy. The afro-futurists, that hack the systems of chronopower and chronography. The speculative realists, that dismantle the barriers to the great outside. The hauntologists, that diagnose the slow cancellation of the future in order to dismantle its enforced depression. (Eshun 2018a: 16:44–17:19)

He passes then five types of accelerationists (untagged, left, right, unconditional, black), four kinds of afrofuturists (untagged, mundane, 2.0, pessimist), three sorts of feminists (cyber, xeno, black poetic) until he arrives even at the Landian 'neo-reactionists, engaged in promoting highly advanced drastic regression' and the:

> Inhumanists, that argue that … inhumanism is a vector of revision that relentlessly revises what it means to be human by removing its supposedly self-evident characteristics, while preserving certain invariances. (Eshun 2018a: 18:43–20:43, passim)

His enumeration concludes with the 'gulf-futurists, that emerge from "the isolation of individuals via technology and wealth and reactionary Islam"', and the 'sinofuturists, that argue that "sinofuturism is an invisible movement – a spectre already embedded into a trillion industrial products – a billion individuals"' (Eshun 2018a: 21:49–22:38). A spectre, I might add, already embodied and thriving in trillions and billions of present, material entities. A spectre, embodied in theories that exceed all fiction written until today. These twenty-five interpretative communities, listed by Eshun, and maybe many more, are actually the way theories are indeed embodied these days. They come into being: their theories will haunt you.

Acid Communism

A new humanity, a new seeing, a new thinking, a new loving: this is the promise of acid communism (Fisher 2018: 767).

Fisher could not finish his programmatic essay 'Acid Communism' that should serve as an introduction to his collected writings. But the torso of this essay was published almost two years after his death. It represents such a possible utopia. A utopia that does not need to be mourned and lamented, remembered and suppressed. This utopia is written or outlined as a joyfully anticipated and eagerly awaited goal one might then actually work for, one might indeed sacrifice a larger amount of one's everyday life for to turn it into a reality. But what is acid communism? Matt Colquhoun (2018) writes:

> In truth, Acid Communism resists definition. The word 'acid' in particular, by invoking industrial chemicals, psychedelics and various sub-genres of dance music, is promiscuous. With so many uses and instantiations in various contexts, it is as difficult to cleanly define as 'communism' is in the 21st century.

Two, if you will, highly suggestive and nowadays seemingly more imaginary entities, *Communism* and *Acid*, two ghosts if you will are actually merged in the concept. Fusioned into one acid communism it might be capable of giving way to a yet unimaginable new hope, new future, new utopia? A utopia that could be outlined with a sentence by Herbert Marcuse, cited by Fisher:

> The spectre of a world that could be free. (Marcuse [1955] 1998: 93, cited in Fisher 2018: 753)

This utopia then requires a conduit to amplify and to multiplex all possible efforts, activities, energies and desires precisely towards this very goal. The concept of acid apparently can serve as this conduit, Matt Colquhoun (2018) claims:

> In this way 'Acid' *is* desire, as corrosive and denaturalising multiplicity, flowing through the multiplicities of communism

itself to create alinguistic feedback loops; an ideological accelerator through which the new and previously unknown might be found in the politics we mistakenly think we already know, reinstantiating a politics to come.

After the bleak lament of Fisher's trilogy *Capitalist Realism, Ghosts of My Life* and *The Weird and the Eerie* since 2009, the concept of acid communism might have led him and it might even lead us, his readers, into thinking of a new, a possible and a better future. A future that is being established from the outside, from the xenosphere, from an alternate thinking and diffracting sensing:

Acid Communism is, then, a project for seeking 'the outside' of sociopolitical hegemony. (Colquhoun 2018)

This is the future. This is the hope. At least of Mark Fisher, maybe of Kodwo Eshun too, but surely not of Nick Land. One might then describe this somewhat self-fulfilling and prolifically strategic progress into a new future – combining several concepts introduced or invented earlier in this book: the mutantexture of a future acid communism is the implex of the mythscientific hyperstition that is articulated in syrrhetic and mixillogic nontologies.

6

NON
Ultrablack Resistance

You sit at a table. You engage in a conversation with others. A lively, vivid, but still civilized, rather calmly moderated conversation on current affairs. The statements go back and forth, the heat rises, then dies down again. Then you say:

> We have the opportunity here to blather like socialists. Some can speak evolutionary talk, others can speak revolutionary talk, yes. And what happens objectively? Oppression will not change at all! Television is an instrument of oppression in this mass society! And that is why it is quite clear here that if anything is going to happen here, one must stand against the oppressor. One has to be biased. This has to be said here. And that's why I'm gonna break this table now. Yes, so that everyone knows! (Nikel Pallat quoted in Steinbach & Szepanski 2017: 85–86;[1] translated by Holger Schulze)

And that's why I'm gonna break this table now. An erratic act occurs. This very act of breaking the table, it breaks at the same time all the carefully established and maintained consensus supposedly supported by everyone – until this very moment. This consensus is now cancelled in the most disruptive and most violent way. It could only have been more violent if one of the participants had violently attacked another one – and thus ending the consensus. But by this performative act of destroying the wooden tabletop – maybe even rehearsed or at least planned ahead before arriving at the television

studio – this very participant is inscribing his dissenting position
unmistakeably into all the other participants' memories and, in
this case, even into the collective memory of all who watched this
TV show, heard or read about it – or watched it more recently in
some online repository. This act took place in a TV talk show aired
on German television on 3 December 1971. The protagonist was
Nikel Pallat, singer and manager of the famous Berlin band Ton
Steine Scherben, well known at the time for popular rock songs
that encouraged and flamed the protest and the unrest of 1968 with
Marxist demands for liberation imagined in lyrics and in melodies.
However, the described scene in the talk show had actually next
to no impact regarding the talk show itself (aside from Pallat,
already on his way out, taking some precious microphones with
him for imprisoned young comrades). Yet, the urge for dissent, for
disruption, for difference, for denouncing and dismantling a false
and fake consensus is archived and memoralized in this act and its
recording.

This setting in of a break, of a separation, of a substantial,
not only occasional disagreement, of a fundamental critique and
disengagement is a dark and a hurtful one. Explicitly stating
such an unbridgeable difference and disruption between one and
another – between me and you – is at the same time painful as it
grants relief. Nikel Pallat's act precisely performs this actual pain
and agony that might have been felt at the time by not so few of
the participants at the talk show's table. Pallat's act externalizes
in his flesh and in his action what was to be experienced as a
repressed affect at the time, three years after 1968. A repression that
might have also materialized in a transition from rather peaceful
activism, demonstrations, interventions of civil disobedience
into meticulously planned, transnational and wilfully cruel and
threatening terrorist acts. But with Pallat's act this agony is, at
least in this very TV show, exposed and not any longer concealed
or covered by polite and gentle small talk. The pain is tangible
in this undoubtedly awkward moment – as it is in any sudden
and eruptive articulation of long repressed feelings. With this
manifestation of pain in an otherwise seemingly all cleaned,
whitened and painless environment for public entertainment,
only with this moment of *breaking the table* the harshness of this
situation could have been exposed. The situation is then thoroughly
blackened. This situation, right now and right here, does not any

longer look as if it were devoid of all characteristics and conflicts. However it turns out to be replete and densely filled with each and every particle of conflict, pigments of taint, contaminated to the fullest with all unresolved struggles of the past decades, in politics and in capitalist exploitation – including the exploitation and degradation of genders, ages, bodies in their abilities, of lifestyles and heritages.

This break from the usual and the dominant, the white, the devoid, blank and hegemonic, this break is *ultrablack*: it is *NON*. This rupture is probably the strongest symptom of dissent. It does not necessarly present or even advertise a better, a more desirable or even possible future utopia. No sense of a desirable *acid communism*, of a *generative syrrhesis* or possible new world is presented as alluring stimulus. Yet, this very act materializes and realizes primarily the existing rupture in lifestyles, in economies, in the *Produktionsverhältnisse*, the conditions of production at present times. NON is a marker of radical and fundamental resistance. NON cuts off the lines of communication and of negotiation with current consumer cultures. NON cuts off the habit of servicing the oppressor, it stops the care work for the oppressor – strictly following Audre Lorde's famous dictum: 'This is an old and primary tool of all oppressors to keep the oppressed occupied with the master's concerns' (1984: 113). This is the ultrablack rupture. A rupture that is performed in radical opposition to an oppressor: 'You have to be biased' (Lorde 1984). For in this radical rupture in all its ultrablackness also stand the twenty-five and more *interpretive communities*, discussed in the previous chapter. But what precisely is an *ultrablackness*?

Ultrablackness

Do you consider resistance against the existing order of things these days necessary? Then where is the foundation for this order in regard to politics, economy, sociality, ecology or even in the sciences and in the arts? How would one then more specifically resist against these – apparently quite questionable yet seemingly rather indispensible – foundations? What constituents, what institutional apparatuses and what interpersonal, what societal and political agreements would then, consequentially, need to be rejected, revised, to be dismantled or bluntly annihilated? For if one would miss only one minuscule but crucial element of this foundation in our habituated order of things, one would surely then never have the slightest chance at all of aiming at a progressive post-institutional arrangement of everyday life. This fundamental critique and resistance against all things present is considered as *black*. This form of fundamental resistance represents therefore a monolithic opposition. This opposition is directed against, as addressed in the previous section, all the conflicts, all the pain, all the suppression, the torture, the violence and the everyday power relations of disciplining and punishing in contemporary societies. The resistance against all of this considers itself as black – and as black it is represented on the political spectrum, in protest marches and in activism. But this black might just not be enough. According to Marxist theorist, head of the record label *Force Inc. Music Works* and chief editor of the online magazine *NON*, Achim Szepanski, the colour of contemporary and future resistance has to be *ultrablack*: a black that is more than black. With black studies scholar Fred Moten one might diagnose here two background effects that might have led to this very notion of the ultrablack.

The first background effect is, what Moten calls *black fugitivity*, a 'predisposition to break the law' (Moten 2003). With Moten this predisposition is grounded in the colonial history of alienated blackness and deportation into the society of a white and alien culture. It is almost necessarily so, that deported persons, not familiar, not educated and neither learned, trained or introduced into all the meticulous details of this alien culture, must get recurrently into conflict with this culture and its arbitrary regulations, laws and etiquettes. This is effectively *not* a predisposition of the deported personnel but of the alien environment they have been deported to. It extends to the children of deported families as they even more so are assigned a life in alienation and already in fugitivity – before they even entered an age or lifestyle of an adult. Now, one could have argued for other forms of fugitivity also, referring to other deported persons, from other areas of the colonized globe, or to other people being alienated for their religious beliefs or other criteria that were chosen at one historical point for discrimination.

However, already the very definition of fugitivity leads to a more specific understanding of blackness in this specimen of resistance. Jack Halberstam defines the concept of *fugitivity* as follows:

> Fugitivity is not only escape, 'exit' as Paolo Virno might put it, or 'exodus' in the terms offered by Hardt and Negri, fugitivity is being separate from settling. It is a being in motion that has learned that 'organizations are obstacles to organising ourselves' (The Invisible Committee in The Coming Insurrection) and that there are spaces and modalities that exist separate from the logical, logistical, the housed and the positioned. Moten and Harney call this mode a 'being together in homelessness' which does not idealize homelessness nor merely metaphorize it. Homelessness is the state of dispossession that we seek and that we embrace: 'Can this being together in homelessness, this interplay of the refusal of what has been refused, this undercommon appositionality, be a place from which emerges neither self-consciousness nor knowledge of the other but an improvisation that proceeds from somewhere on the other side of an unasked question?' I think this is what Jay-Z and Kanye West (another collaborative unit of study) call 'no church in the wild.' (Jack Halberstam in Harney & Moten 2013: 11)

Fugitivity is being separate from settling: being together in homelessness, the state of dispossession that we seek and that we embrace. Fugitivity therefore represents a starting point of resistance that first of all accepts its own dispossession, its being discarded and disavowed. This *undercommon appositionality* represents a constant state of being alienated: a state from which any activity of resistance, of demanding and of building another framework of social, economic or cultural life can only set in. To set this as the ground of blackness and ultrablackness frames it in a way that doesn't ignore the *sonic colour line* (Stoever 2016) reconfirmed here but that reflects upon it and results in acknowledging the involuntary trajectory towards fugitivity, towards questioning the social commons, and hence also working towards a revolutionary state. A major part of this revolutionary strategy of blackness and ultrablackness is then the instalment of a fundamental rejection of existing institutional procedures, regulations and orders of speech, of ideological frameworks, of epistemological and ontological, of ethical routines and habitual forms of behaviour. Everything has to be rejected – and everything has to be rethought. It needs an act of major and painful disruption, an act of distancing, maybe even of violently marking a break, a rupture, a stopping of routine communication: the axe that is hacked into the table during conversation. Szepanski recognizes precisely this distancing and disruption in the theory practice of sonic fiction. For him, sonic fiction is a striking approach and research strategy that acknowledges, scrutinizes and acts accordingly to this catastrophe. An almost necessary catastrophe as he suggests:

> It is a kind of disaster studies, an act that breaks down the formal structures of space and time. In the mimicry of this approach to electronic music, both in science and in music, the formal structures of time collapse, regress to mud, and space is pushed back and forth until it bends to be trampled by the pulsations of alien music, while the thinking space becomes seasick. (Steinbach & Szepanski 2017: 66;[2] translated by Holger Schulze)

The formal structures of time collapse, regress to mud, and space is pushed back and forth until it bends to be trampled by the pulsations of alien music, while the thinking space becomes seasick: this is the disruption as it can be experienced when ultrablackness hits you. When the all-consuming, all-absorbing and all-imploding might of ultrablackness exercises its power of radical, pervasive and fundamental negation. The one message, the one action, the one intervention of ultrablackness is taking an axe and ramming it into the fake common ground or shared table and says: NO.

The critical step here is the construction of the exclusive opposite. Underground Resistance say somewhere that disappearance is our future, and according to Eshun the Black Power of UR should therefore be invisible, not identifiable, hidden, unrecognizable and not public. (Steinbach & Szepanski 2017: 71;[3] translated by Holger Schulze)

This unidentifiable blackness of the exclusive opposite is represented by ultrablackness. The sonic warfare performed, enacted and facilitated by artists, designers and musicians is unhidden as a revolting practice:

> With this, the war machine is completely ready for action. As a mob of machinists with technical apparatuses, the guerrilla unit fights against the machinic-urban machine body of capital. Writing and music can also be war machines. (Steinbach & Szepanski 2017: 74–75;[4] translated by Holger Schulze)

NON

The radical restart of a revolutionary disruption in the sciences and humanities then sets in for Szepanski with the non-philosophy of François Laruelle. Laruelle's project of a *non-philosophy* is driven by comparable desires as they have been articulated by Michel Serres, Gilles Deleuze and Félix Guattari or Brian Massumi. All these authors and thinkers come to the painful diagnosis that the long tradition of academic philosophy has actually left all the pressing and actual issues of thinking, sensing, living and doubting in everyday life and in the individual struggle of existence smugly behind. Such an attack against academic philosophy is, obviously, one of the most noble rhetoric figures in the history and in the arsenal of philosophy itself. It represents, again and again, the overly excited signal horn and fanfare before introducing yet another branch of philosophy that will be assimilated, sooner or later, into academia. In this case here, though, this urge to bring the untamed, the wild and unordered, the chaos of sensation and desires, of practices and sensibilities, of fears and of the real back into a discursive format of thinking actually *did* generate and still generates radically new forms of writing, of conceptualizing, and even of sensing. However, it also contributes to the establishing of new forms of academic philosophy – which needs to be considered a major flaw, in this case, alas; I will come back to this paradox later in this chapter.

This *non-philosophy* – and equally any kind of *non-studies, non-science* or *non-research* – does not actually signal the end or a prohibition of scholarly research activity. The prefix or epithet *non* more specifically puts a halt to one common, rather unquestioned and fundamental practice in research: the practice of excavating and then distinguishing dialectically the fundamental and ontological categories in, what researchers call, *The World* or *The Reality*. Such first distinctions, enshrined in the notorious, scholarly phrase 'Draw a distinction!' (famously ascribed to mathematician George Spencer Brown, and later almost compulsively cited by German systems theorist Niklas Luhmann) apparently are the foundation of any endeavour, any effort, any project or argument in modern scholarship. Whereas such distinctions seem to be fundamental to the established habits of scholarly research and academic writing, they can effectively hide some more contestable if not questionable insights, findings or *worldviews, Weltanschauungen,* in exactly these unquestioned axioms before drawing the first distinction. It is a well-known sleight of hand in theory and philosophy to hide all the potentially most contested propositions or claims precisely in the prerequisites *before* the first distinction – so not many readers might even question those.

In contrast and in resistance against this practice of hiding crucial interpretations a non-philosophy following Laruelle goes back to this very start. In this very respect it is neither an *anti-* nor a *meta*-philosophy. Both interpretations can be found in the history of philosophy – each time implying and operating carefully with yet just another major distinction or a series thereof – for example taking the *body* or *perception* or *energies* or *infinite regress* or *continuous processuality* as a prerequisite for all subsequent distinctions, categorizations, reflections and arguments. So, if one would translate non-philosphy's effort as: *We must go back to the things themselves!* then this would only recall the efforts of phenomenology since its beginnings around 1900: 'Wir wollen auf die "Sachen selbst" zurückgehen' (Husserl [1913/21] 1984: 10). The efforts in this case have indeed primarily achieved the establishment of a new branch, a new school, and a number of new chairs, departments, academic societies, journals, handbooks and encyclopaediae along its path of doing philosophy in a new way, *phenomenologically.* Yet, following Laruelle and other non-philosophers this precisely is what they do not intend to achieve. A non-philosophy to the contrary does continually reject any operating within the philosophical realm of a spectrum of interpretations, branching out ever further and further, into ever more detailed differentiations and distinctions, ever more departments, journals and handbooks. These instalments though could be, from one point in time onwards, a possible side effect of this research activity but it surely is not the first desired goal of non-philosophy. 'Decisional closure' (Tilford 2017: 140) is being infinitely postponed. If a non-philosophy now indeed makes a strong effort *not* to become just another part of the scholarly philosophical canon, what then is it? How can one operate and not be quickly assimilated or find oneself drifting into a new branch of philosophy? Or, more pragmatically: what kind of activity – if it is not drawing distinctions, arguing, proving and reflecting – are non-philosophers then actually performing? *What are non-philosophers?* François Laruelle gives the following answer in the form of an almost literary description of non-philosophers:

> I see non-philosophers in several different ways. I see them, inevitably, as subjects of the university, as is required by worldly life, but above all as related to three fundamental human types. They are related to the analyst and the political militant, obviously, since non-philosophy is close to psychoanalysis and Marxism – it transforms the subject by transforming instances of philosophy. But they are also related to what I would call the 'spiritual' type – which it is imperative not to confuse with 'spiritualist'. The spiritual are not spiritualists. They are the great destroyers of the forces of philosophy and the state, which band together in the name of order and conformity. The spiritual haunt the margins of philosophy, Gnosticism, mysticism, and even of institutional religion and politics. The spiritual are not just abstract, quietist mystics; they are for the world. This is why a quiet discipline is not sufficient, because man is implicated in the world as the presupposed that determines it. Thus, non-philosophy is also related to Gnosticism and science-fiction; it answers their fundamental question – which is not at all philosophy's primary concern – 'Should humanity be saved? And how?' And it is also close to spiritual revolutionaries such as Müntzer and certain mystics who skirted heresy. When all is said and done, is non-philosophy anything other than the chance for an effective utopia? (Laruelle 2004)

Non-philosophy is, therefore, closer to activism, to a pragmatic lifestyle that intends to transform everyday life, maybe the arts and sciences, but surely politics and public discourse: *should humanity be saved? And how?* Non-philosophy asks the questions of *science fiction* in *science's reality*. Again in this case the imaginative, suggestive and viral potential of science fiction is taken up and employed to energize and to accelerate the effects of an intellectual practice. The literary richness and elaborated details, the realistic complexity of issues and their possibilities of manifesting in science fiction and its thought experiments make it also here a very versatile genre to enhance even, against all clichés, the realism, the pragmatism and the activism of a theory. This theory practice of non-philosophy, is – and this brings it very close to the writings of Serres and Eshun, also of Badiou or Debray – characterized fundamentally by its *radical immanence*. Non-philosophy does not primarily intend to erect yet another interpretative edifice on top of other insights, research findings and artefacts and 'to manufacture artificial problems to suit its own pre-determined and ideologically imposed solutions' (Tilford 2017: 140); to the contrary, the goal of this new and non-standard philosophy is to describe and to analyse the immanent reality indeed in its immanence. Non-philosophy rejects therefore traditional philosophy's,

> Peculiar arrogance toward its object of inquiry … the pretension of philosophy to elevate itself above any object or discourse in order to offer a philosophy *of* it: a philosophy *of* science, *of* art, *of* music, etc. … Convinced that its object is fundamentally ignorant about itself, philosophy is little concerned with what that object has to say on its own behalf. (Cox 2013)

Non-philosophy, hence, sides with its *object of inquiry* and assumes it as the actual and leading *subject of inquiry*. Non-philosophy is in this very sense non- and unphilosophical. It does not elevate itself above its objects, yet it thinks with the practice theory inherent to the subject in question: it explores what this research subject has to say on its own behalf. Should one call this then a *barbarian philosophy* in the best sense? Or, when inverting an insult into a noble epitheton: a *crude philosophy*? Because a non-philosopher rejects thereby any actual separation between certain crafts, sciences or practices on the one side of everyday life – and their reflection, modelling or theoretization on the other side of an academic hermitage. Non-philosophy recognizes that such a separation is factually and methodologically an ancient and dead distinction that bears next to no reference in the reality of commodified research activities acting in a market of funding bodies. Non-philosophy claims that the best and most appropriate description of a field in reality emerges almost inevitably, *generically* out of the intrinsic actions, operations and reflections present in this very field that is being reflected, indeed an *effective utopia*. Achim Szepanski, probably one of the most energetic and rigorous agents of non-philosophy these days in the German language, translates this non-philosophy then into a *non-musicology* – enmeshed with Kodwo Eshun's approach of sonic fiction:

> Sonic thinking or non-musicology composes theory as its own object, writes an autonomous music fiction .... Fiction implies performance, invention, artifact and construction, not only in a non-expressive and non-representational sense, but rather as immanence. (Steinbach & Szepanski 2017: 63;[5] translated by Holger Schulze)

Sonic thinking as non-musicology emerges, according to Szepanski, from sound practices and sonic artefacts. It is *not* separated from them but is their adequate continuation (Fowler 2015; Steinbach & Szepanski 2017). The craft of musicking and sonicking is not distinct from the craft of theorizing, investigating, or analysing sonics and musics. Non-philosophy insistently demands *not* to separate the research practices of a discursive reflection from precisely those particular material practices they are actually reflecting on. Following Laruelle, this non-separation is even one of the main characteristics of a non-discipline. This 'non-decisional immanence' (Tilford 2017: 141) though does not entail that there could never exist separate fields of practice or of social experiences. Yet, this non-separation implies that any scholarly approach to a field of practice has first to acknowledge the generic, the field-specific and the materially informed discourse and reflections grounded in the practice of the field in question. Scholarly reflection, then, must *not* open up a new and superior (and often detached, patronizing and condescending) discourse outside this field of practice. A non-discipline is grounded and based in these practices and their generic forms of knowledge. Again, Eshun's famous dictum resonates here with a strong repercussion: 'music today is already more conceptual than at any point this century' (Eshun 1998: -004—003) Or, again in Michel Serres's concepts, Syrrhesis trumps analysis.

Rhythmight

For Szepanski and his approach to sound the *rhythmight* is essential. Like in other areas of non- or de-disciplinization following François Laruelle's non-philosophy, also the non-discipline of sound and music revises established idealist and historically tainted concepts – tainted by strategic and political concepts of *The Human*, of *Freedom*, of *Wealth*, of *Liberation,* or of *Democracy*. This materialist revision is, obviously, not a merely academic endeavour. As Terry Eagleton recently pointed out, it is an epistemological urge rooted in the fact that 'the senses ... are constitutive features of human practice, modes of engagement with the world' (2016: 62) – and at the same time, following Klaus Theweleit (2018): 'People live in bodies, people are bodies.' This new materialism then in all its manifold varieties, be it as sensory, as sonic or as feminist materialism, then seems to ground contemporary and progressive research in a more fundamental sense.

Non-philosophy and *non-musicology* undoubtedly follow this path, yet without the common pathos or fervour often to be found in more recent approaches to materialism. Instead, non-musicology and non-philosophy both undercut and undermine contemporary edifices of theories as they are taught in universities and regarded as *common sense* – in the worst, most ideologically repressing sense of this word. Programmatically, this new musicology, *non-musicology* or *Musicology?*, with a question mark significantly added to the word, starts out with the physiological and the material substance of sound and listening at its core as postulated by composer, performer and producer Jarrod Fowler (2015):

> Musicology? [*sic*] is the proper start of non-musicology nearly freed from the vicious circle of the musical composition. Musicology? strikes from samples to pulse 'rhythmights' with rudiments through an anticausal method of percussive dialetheics (rhythmics), which (inconsistently) counter-counts.

Such an explication of non-musicology – cited by Steinbach Szepanski in *Ultrablack of Music* (2017) – is primarily referring to musical and sonic substances, their physical and physiological effects as well as their complex meshwork made out of affects and reflections. Eshun is following with sonic fiction such an explication. Experiential phenomena are being explored and narrated here that probably hadn't been thought of by listeners, critics or even researchers for quite a while – yet, they constitute possibly the major object of reflection for generic practitioners, producers and musicians. The material and affective substance of sound, the oscillations, amplitudes, the swinging and the percussive rhythms are inextricably melted together in this generic approach by assuming a deeply monist *hearing perspective* (Auinger & Odland 2007). Rhythmight, hence, is one of the typically generic categories a non-discipline such as *non-musicology* would propose:

A rhythmight is the non-musicological term for a non-musical practice of indifferent hearing that replaces the formerly narcissistic music of X. (Fowler 2015)

With this definition the detour from an anthropocentric focus on individual expressions and a dynastic genealogy of artists, schools, styles and music history is performed. This pervasive genealogy is replaced instead by a transhuman, a situated, and a relational category; a category encompassing all sorts of agents, intensities and effects. Fowler (2015) writes further:

So, for instance, instead of some 'music of science' or some 'music of philosophy' one unstably pulses with non-musical axioms some rhythmight of music and science or some rhythmight of music and philosophy. A rhythmight is unstable because the theory is also only occasional, such that non-musicology's axioms are damped, but the practice of the theory is utterly dependent on the samples available and revisable upon the availability of new samples. In both cases the rudiments retain their counts: neither music, science, nor philosophy is subsumed within the other, because the rhythmight is constructed from the axiomatic anticausality of Rhythm as counter-counted and the method of Rhythmics that hears from rhythmicity.

Neither music, science, nor philosophy is subsumed within the other: they are constructed as mutually dependent and interpenetrating each other, emerging out of a monist practice of *axiomatic anticausality* – all in tune with Laruelle's non-standard philosophy. Like the practice of sonic fiction also the monist term of *rhythmight* is developed out of the experienced and generic practice of musicians, of listeners, and of composers. It is *not* developed out of the urge to construct a superior and secondary, an external and thoroughly consistent and self-indulgent system of thoughts, definitions or recursive definitory processes to explain all musics on all occasions and all situations, in all contextual and cultural and societal frameworks. This rather obsessive compulsive desire, I might add, of nineteenth-century research – driven to a large part by imperialist and territorialist desires – is abolished by non-philosophy. It gets also abolished by theorists and musicians such as Jarrod Fowler or Achim Szepanski who put thereby the heuristics of sonic fiction into practice. As a consequence, Szepanski's request resembles here indeed a sort of renaissance of musicology – as non-musicology:

> Non-musicology by no means demands a new musicology, but a generic science of music, or to put it another way, not a science, but rather a heresy or fiction in the face of music. (Steinbach & Szepanski 2017: 62;[6] translated by Holger Schulze)

This heretic scholarship is precisely what *More Brilliant than the Sun* proposes. It is a generic exploration of sounds and performativity that can be found in the articles, performances, lectures, artworks and books that continue to work with mixologies, mutantextures and mythsciences. In the process of writing and thus contributing to sonic fiction, these concepts foundational to this *heresy or fiction in the face of music* might seem at first radically erratic and inconsistent; and yet, they are actually more consistent to musical practices and sonic experiences than the vast amount of historically established conceptualizations that are factually external to musicking and sonicking. The concepts emerging from sonic fiction are generic to sound and music – and that is their primary quality. The concept of rhythmight is such a generic concept because it puts a monist understanding of rhythm and hearing at its centre:

The interaction between Rhythm and hearing is unilateral because the relationship only goes one way, hearing cannot affect Rhythm, Rhythm is foreclosed to hearing. Non-musicology radicalizes this notion by subtracting hearing from the framework of experimental music and setting hearing within an exological realist framework where Rhythm is what is unilaterally anticausal, without that then anything is simply reduced to Rhythm, but rather everything music claims to master is heard from Rhythm. This axiomatic description of the anticausal interaction of Rhythm with hearing frees music from the pretence that music can adequately listen to Rhythm, this is music's condition of negative freedom, while at the same time freeing music to hear inadequately, that is non-musically, the various fulcrums, we may even say 'silences', of Rhythm that experimental music has concerned itself with, experimental music's condition of positive freedom. (Fowler 2015)

One could now claim that rhythm is here transformed into a new unquestionable axiom, maybe even a new metaphysical foundation of musicology. And indeed, this very heresy can be considered its most noble goal:

Although Non-musicology is critical of musical doctrine, Non-musicology does not goad the absolute destruction of music, but strikes some unknown invention of music. The program of Non-musicology is to use musicology to construct alien theories without those theories being yielded by the Principle of Musical Sufficiency: 'All is not musical, this is our news.' (Fowler 2015)

Sonic fictions precisely are these alien theories of sounding, receiving, transmitting and experiencing sonics and musics. It imagines, conceptualizes and builds the *futurhythmachines* (Eshun 1998: -010–009) of which Eshun writes. This non-musicology reacts to a non-music, a radical music as described by Szepanski:

Radical music resembles a kind of black box: it is a music box of and for blackness, and the thinker and the consumer of music take a place in the black box themselves and do not approach the box from the outside. There is a non-musical triangularity to report: The (multiple) producer who lets the transversality of the black be sounded; the black jukebox as an infinite sounding of the incomprehensible/black; the consumer who hears extracts of infinity in the black jukebox. (Steinbach & Szepanski 2017: 69;[7] translated by Holger Schulze)

The rhythmight in this non-music of ultrablackness requires then an adequate non-musicology, capable of exploring and scrutinizing the generativity and the effects of sonic artefacts. These generic artefacts then, in their radical negation and rejection presented as monist rhythmight, are constituting and promoting an ultrablack resistance. Rhythmight is, if you will, the resisting substance of material affects that enables ultrablack performers to act in sound.

Ultrablack Resistance

The extremist endpoint of this resistance – that can be traced back to ultrablackness, as this chapter showed – is a sensory practice that is activism and aesthetics at the same time. It can be represented by the very moment – introduced at the beginning of this chapter and included in Steinbach and Szepanski's book on ultrablackness (2017: 83–86) – when Nikel Pallat axes a table at a talk show on public television. This moment can and maybe must be regarded as a *non-musical* action. Definitely, it constitutes an object of non-musicology; but also an act of corporeal resistance, without doubt. This becomes very apparent when Paul C. Jasen writes towards the end of his *Low End Theory* the physio-logics of three *bass cults* Jungle, Dubstep and Footwork:

The aim of jungle's breakbeat science is a body in flight, or maybe more accurately, a molecular body pulled out of itself along multiple, fractal trajectories by the heterogeneous momentums of its broken breaks. (Jasen 2016: 178)

Toothy, envelope-filtered pulses play a slow back and forth with a cleaner, heavier sub. One rises roughly out of the chest and smacks against the walls, the other is more barometric, weighing on the room as a whole …. In this new rhythmachine [of Dubstep], it was the undulations of multilayered basslines, rather than the beat, in any familiar sense, that gave the physiologic its rhythmic texture. (Jasen 2016: 180)

Footwork is a competitive dance culture .... Here, sonic body and rhythmachine become difficult to separate, as blurring feet under strangely still torsos become the rhythm that seems to be missing-but-implied in the stripped-back tracks. (Jasen 2016: 182)

In these three steps Jasen contributes to a non-musicology of electronic dance music. The three dance practices are constituted by a bass materialism of corporeal practices that organize the bodies of its dancers as well as the sono-machinic generators of their beats, sound events and bass lines. The transformation of these examples of non-music as well as the guidance and control they exert on dancers represents precisely the aforementioned amalgamation of aesthetics and activism. These dances and bass cults are resistance as they are ultrablack:

All music was a variation of the human machine interface. Suddenly sound machines were just as cyborg as gigantic corporate simulations. (Eshun quoted in Weelden 1999)

They are not entertaining or occupying your supposedly *disinterested appreciation*, your *interesseloses Wohlgefallen,* as traditional Kantian aesthetics would have demanded. They are indeed and more directly, technopoetically and in siturelational effects, exerting sonocorporeal control. This control operates radically colourless, in *uchromia* as Laruelle calls it, in black:

Our uchromia: to learn to think from the point of view of Black as what determines color in the last instance rather than what limits it. (Laruelle 1991: 3)

From this ultrablack *non-decisional immanence* of sound the effects of *black technopoetics* (Chude-Sokei 2016) begin to unfold and to expand. A diffracted non-musicology can now set in – maybe even as another, convincing and radical example of decolonizing research – following the famous reminder by Audre Lorde:

For the master's tools will never dismantle the master's house.
(1984: 112)

These new and decolonized tools of non-music and of non-musicology indeed complete one of the main projects of the CCRU in Warwick to which at least some figures of thought and pervasive desires in Eshun's writing can be traced back. This non-musicology in alliance with Eshun's sonic fiction is indeed – citing Simon Reynolds's (2009) description of CCRU's approach:

Theory melded with fiction, philosophy cross-contaminated by natural sciences (neurology, bacteriology, thermodynamics, metallurgy, chaos and complexity theory, connectionism). It's a project of monstrous ambition. And that's before you take into account the most daring deterritorialisation of all – crossing the thin line between reason and unreason. But as they say, later for that.

This monstrous project quite consistently converges with François Laruelle's non-philosophical project; the latter though more widely discussed and indeed performed and adapted, the former yet still in its brief period of existence more lastingly materialized and institutionalized. Non-musicology, understood as a sort of reinvented musicology in the tradition of the CCRU, returns to the ultrablack resisting immanence of sound:

See black! Not that all your suns have fallen – they have since reappeared, only slightly dimmer – but Black is the 'color' that falls eternally from the Universe onto your Earth. (Laruelle 1991: 4)

Inconclusion
Six Heuristics for Critique and Activism

Heliocentric effect: rising spirals of analogue synth, cascades of keyboard runs, the rings of Saturn, a female impersonation of the Siren from outer space you know the sort of thing.

(Eshun 1992a: 66)

Coming from outer space into the precinct of sonic fiction you had the chance to encounter in this book a truly erratic series of interpretations, appropriations, creative misreadings and deformations of what might have been in the mind of the humanoid alien that carries the name of Kodwo Eshun on the original book cover of *More Brilliant than the Sun* from 1998. Hence, there are almost no precise definitions of sonic fiction to be found. Eshun gives in his writings only rarely explications that are then even more of an implicit and inductive kind than being actually explicit and deductive: Eshun shows what sonic fictions are by writing them. The more traditional hermeneutic way to excavate and to distillate precise definitions from contaminated and blurred sources, therefore, didn't lead very far. Eshun's writings are just as unstable and oscillating in their conceptualizations of sonic fiction as any later approximations, appropriations, deformations or reformations.

Therefore I chose another way for understanding this concept – not only by going to the original or supposedly uncontaminated sources. I tried to understand this concept in a kind of combined

futurological and archeaological triangulation, using precisely
these manifold and conflicting interpretations and applications as
the only material references and traces to work with. In the midst
of the heated conflicts between interpretations and realizations
of sonic fiction, I assumed, at least some recurrently addressed
constituents of sonic fiction would become obvious – as well as the
influence and the impact of some more idiosyncratic and strange,
yet widely accepted perspectives. These stable constituents and
dubious outliers might then provide at least the ledger lines for
a tentative sketch to answer the question: *What is sonic fiction?*
At this point of my investigation I would therefore like to collect
all these remnants and insights granted by all the approaches
discussed and interrogated, flipped over, scrutinized, traced back
and probed in the previous six chapters. The simple question in
the title of the first, introductory, or, more precisely: *extraditory*
chapter of this book will then, again, be the invisible headline of
this conclusion or, more precisely: *inconclusion* to this book: What
is sonic fiction?

Sonic fiction is not a finite genre or method or approach. Under
this moniker a wide variety of academic and non-academic, artistic
and non-artistic, musical and non-musical endeavours can be
summoned. Sonic fiction is a generative and mixillogic, a syrrhetic
and generic, a mythscientific, nontological and decolontogical, a
mutantextural, implectural and multiplestomological endeavour –
propelling its readers, writers and thinkers, its sensors towards
an acid communism and ultrablack resistance of NON. However,
depending on the more precise area of expertise and enquiry in
which a writer, artist, researcher, musician, a thinker or a public
intellectual activates this ferment of sonic fiction, it will react,
connect and coalesce differently with neighbouring entities,
substances, processes and practices. It might then, indeed, generate
quite different and ever more changing, altering and surprisingly
transforming effects in result. Sonic fiction is a highly reactive
ferment in any artistic, everyday or research activity.

There are probably as many emanations, materializations,
mutantextures and interpretations of sonic fictions as there are
soft machines like you and me, taking this idea, its immediate
and intuitive aura as a fascinating inspiration and trigger to do
something with it. In this book I gathered six approaches that

seemed to me representative for the recurring motifs, the trace elements as well as erratic and generative misreadings that can be found in research, the arts, in literature and theory of the last twenty years. Some of the approaches I discussed are more erratic and singular, others are represented in a larger cluster of examples, some are only loosely connected to each other and to Eshun's concept – others are engaging in artistic and theoretical exchange with each other, referencing, questioning, thinking with each other's interpretations and applications. Three of the approaches discussed in this book – in chapters 2, 4 and 6 – refer to everyday practices in social activism, in research and in political activism that can be altered directly by the infusion with sonic fictions; another three approaches scrutinized here – in chapters 1, 3 and 5 – are effectively methods of critique that refer to sonic artefacts and how to infuse them with or distil from them the ferment of sonic fictions. Sonic fictions provide an infusion for activism either by facilitating and accelerating it or by substantiating and mutating it by its prolific forms of critique. Sonic fictions are *heuristic fictions* in this sense (Schulze 2005: 17–19; Vaihinger 1913). They alter reality: they motivate lazy and often self-indulgent humanoid aliens like you or me to take action.

As a heuristic fiction then sonic fiction can be able to assume primarily two roles; these roles are complementarily related, relating to, and relying on each other: the more *activist approaches to sonic fiction* take action directly in the area of the political, the social and the epistemological *through sound*; the more *critical approaches to sound studies*, however, activate in sounding artefacts and imaginations the inherent potential to alter precisely these areas of the epistemological, the social and the political. Both specimens can be found in the population of sonic fictions – and both specimens materialize pervasively the critical relation between the realms of the sonic, the sensory, the material and the realms of fiction, of imagination, of meaning. In this final chapter, this *inconclusion* at the end, I will now explore the individual characteristic traits of these two specimens of sonic fiction and how they differ from or resemble each other when interpretating and applying sonic fiction. This exploration should give you, the reader, a more concrete idea of how you might be applying, thinking and working with, working through sonic fiction in your current endeavour.

Sonic Fiction as Activism

Sonic fictions operate in the area of the political and the institutional. Sonic fictions inspire activists to use sound, sound environments, sound events and sound practices as a means of political resistance (cf. Chapter 6 in this book), a means of transforming epistemologies (cf. Chapter 4) and a means for social progress (cf. Chapter 2). Through all these areas sonic fictions can alter this present world and its predominant imaginations and fictions. Sonic fictions take action and intervene.

Sonic resistance – as explored in Chapter 6 under the title of 'NON' – is probably the most direct expression of activism enabled and fostered by sonic fiction. Sonic fiction as one core element in non-musicology and its explorations of the rhythmight leads to a thorough recalibration of the function, the meaning and the impact of sounds and music: sonic resistance mutates musical aesthetics to become a force in everyday life and in political struggles. Sonic resistance constitutes a critical if not revolutionary force that does not limit its effects to a stage, a concert hall or a dance floor. This non-music might not necessarily be a sonic weapon in the technical sense, it also surely is not restricted to provide an aesthetic experience alone – but it facilitates an instrument that can be used to transform social, economical and political relations – as well as psychological, biological and sensorial constitutents of you and me, humanoid aliens being sonic personae. In this respect the resignification of the sonic as a form of ultrablack resistance that authors such as Szepanski, Fowler or also Jasen propose, is actually an acknowledgement of music's and sound's painful impact on society and culture – without turning this insight into yet just another branch of sociology, cultural research or musicology. A sonic intervention with this background is foremost a form of political activism. This activism however is driven by a painful sonic critique.

A *sonic epistemology* – as explored in Chapter 4, 'Sensory Epistemologies' – is the foundation for such a re-evaluation of the function of sound and the senses through sonic fiction. This new impact relies on a rehabilitation of a research strategy that values primarily a surprising synthesis, a creative generation of a new artefact as research – being a truly experimental example of a *syrrhesis fiction*: a concept that refers to Michel Serres's thoughts in his book *The Five Senses*. Instead of foremost acknowledging the research value of analytical methods scrutinizing a given artefact or found material by a supposedly anonymous research agent, here the researchers and their material bodies, histories, idiosyncrasies and skills are generating with a *syrrhetical method* ever more new artefacts and new amalgamations of materials: knowledge and insight is therein generated in the very process of performing such a syrrhesis – combining and mixing, remixing and kneading, fusioning and interweaving existing materials and sources and substances. Generativity – yet not necessarily the production of new objects, products, commodities – is inherent to this epistemology. Such a practical epistemology or an epistemology of praxis hence generates a potentially endless sequence of multiple, malleable and new epistemologies. The recombinatory possibilities of generating knowledge through practices applied to materials by specific bodies of researchers and their sonic personae are, supposedly, as manifold as the practices themselves. Each everyday activity, practice and craft that performs a syrrhesis of fusioning or mixing, of performing mixillogics must consequentially be regarded as a potential sensory practice at the core of a new and generative epistemology. These new epistemologies are, after all, apparently fuelled by sonic sensibilities in all their erratic and possibly disturbing richness.

Social progress – as the main goal in Chapter 2, 'Social Progress' – is at the centre of both aforementioned applications of sonic fiction, be it sonic resistance or sonic epistemology. A sensory or sonic practice in the field of the social must be regarded as the core action being taken as soon as an implex – following the conceptual explications by Dath and Kirchner on the ground of Valéry's initial reflections – is emerging and materializing as a new mutantexture in the course of working with sonic fictions. As soon as the dialectics of the implex are in effect, then the sensibilities of researchers and activists are capable of turning into reality what previously might have been thought of only as a strange and improbable future vision. The ground for this material intervention and implementation of improbable ideas into everyday life, experience and commerce are specific sensory practices: these practices transform their agents as well as they transform the material sensory constellations present or the concepts of the sensory as such. Social progress then actually takes place through such transformations of the senses and sensory cultures and their effects and materialization in social interactions and interpenetrations. This progress implies, therefore, a good amount of sensory and sonic thinking.

In these three examples of activism, sonic fiction can and is, apparently, being applied. Sonic fiction as activism is never unrelated or detached from critique – yet is relying on and employing it. It is activism through sensibilities and criticism as it is an activism through practices and actions. This at least is what the critics, researchers, authors, performers, musicians and artists discussed in the relevant chapters of this present book did make until now of Eshun's original concept – from Achim Szepanski, Jarrod Fowler over Paul C. Jasen to Dietmar Dath and Barbara Kirchner, even non-genealogically sidestepping to dancers and performers of Jungle, Dubstep and Footwork, to Nikel Pallat and Ton Steine Scherben, Michel Serres, François Laruelle, Henri Lefebvre, Audre Lorde, Paul Valéry and Underground Resistance.

Sonic Fiction as Critique

Sonic fictions operate in the zone of sensory imagination and theories. Sonic fictions enable critics and writers to use imaginary worlds, theoretical fictions and generative concepts by thinking sonically in general (cf. Chapter 1 in this book), by performing a critical decolonization of sound (cf. Chapter 3) and by further developing tangible utopian scenarios (cf. Chapter 5). Through the use and the imagination of sound, sonic fictions materialize, refine and alter the range of worlds possible for us. Sonic fictions perform critique and method.

The *sonic sensibilities* – as explored in Chapter 5, 'Acid Communism' – that are activated through sonic fiction alter the perspective on affects, experience and social interpenetration. The retronostalgic desire and melancholia for a lost utopia that Mark Fisher excavated in his writings transforms the understanding of political developments in contemporary societies. Unlike other possible sonic fictions of a timeless, ahistorical utopia or of infinite futurist progress, this interpretation stresses the sadness and also the distinct hopelessness that is felt in recent cultural production; a hopelessness that is so radical and so ubiquitous, and that seems so invincible that only another radical step, a shift, a radical leap seems to be capable of altering anything. The sonic sensibilities challenged by sonic fiction therefore are – through all their bleakness in this specific case – motivating and driving forces. These sensibilities are never self-indulgent. They represent, first of all, how the anticipation and compulsion that, apparently, can be understood as the ghosts of our present times that lead to theories that are actually performed, lived and embodied. These sensorially anchored theories then – hopefully – might lead a way out of the impasse and into an imagined future that could be called by the name of Acid Communism; surely, in a thorough resignification and reinterpretation of both concepts connected therein, *acid* and *communism*. The driving sonic sensibilities though are themselves the highly energetic core that is activated, kick-started and employed to support and to drive subsequent actions and interventions leading towards this goal. This goal is social progress.

The theories though, that are embodied here, they represent a differing, however immensely prolific, kind of thinking. This *sonic thinking* – as explored in the very first chapter of this book – proceeds foremost along the lines of a mixillogic mythscience of mutantextures. According to Eshun these three concepts represent core characteristics of sonic fiction: the mythscience of a sonically exerted violence, as Steve Goodman explicated it in *Sonic Warfare* (2010); the rather deviant and alternative *mixillogics* of the epistemologies inherent to *sonic materialism* (Cox 2011; Voegelin 2012; Cobussen, Schulze & Meelberg 2013; Voegelin 2014; Schrimshaw 2015; Lavender 2017; Thompson 2017; Cox 2018), opening up manifold sensorial and logical operations differing from other epistemologies; and, finally the emerging *mutantextures* of *Sonic Possible Worlds* (2014) that Salomé Voegelin explored through a large number of artistic works of sound art, sound performances, media and radio pieces. With these three constituents it becomes possible to define sonic thinking through its form of knowledge, its operating logic and its tangible, resulting artefact: sonic thinking exhausts the reservoir of mythscience through its operating mixillogics and generates thereby a hitherto unknown mutantexture. This digressive yet highly generative character of sonic thinking makes it one of the most apt approaches to analytically approximate a given sonic environment, a sonic artefact or certain observed sound practices. Sonic thinking, therefore, is the critical and prolific method of sonic fiction to analyse, to scrutinize and to understand the sonic. This new understanding then effects and triggers directly the new sonic epistemologies.

This main *sonic critique* – as explored in Chapter 3 on 'Black Aurality' – becomes crucial if not revolutionary when indeed sensibilities and thinking through and with the sonic are performed. At this point the diffracted sensing and thinking with the sonic – following Barad's concept of diffraction – generates through aberrant yet instructing autohistories – following Anzaldúa – and nontologies the new knowledge and the awareness, for instance, of a *black aurality*. From this starting point of critique a multiplicity of more, probably intersectionally informed auralities – of *AlterDestiny* (Sun Ra) and an *Alter Nation* (Eshun) – can be unfolded, further developed, applied, exemplified and put into action. Sonic critique is, hence, not restricted to a critical analysis of sonic artefacts alone, providing dissenting sensibilities in thinking, but transgresses into the actual framework for taking political and institutional action. Sonic critique effectively is a sensory critique drawing the critic almost involuntarily into the whole of the political meshwork of historical discrimination, power strategies as well as into contemporary struggles and efforts towards a liberation of this variety of sensibilities: decolontologies in action. These specifically sonic sensibilities and forms of thinking generate then new specimens of critique and practice in the realm of the sonic that are fundamentally operative to foster ever more new sonic epistemologies and to provide the means and the grounds for sonic resistance.

In these three examples of critique, sonic fiction can be and is, apparently, being applied. Sonic fiction as critique is, as is obvious, never unrelated or detached from activism – yet facilitating and promoting it. It is a true theory of practices and of action as it is a theory of sensibilities and criticism. This at least is what the critics, researchers, authors, performers, musicians and artists discussed in the relevant chapters of this present book did make until now of Eshun's original concept – from Mark Fisher, Boards of Canada and Brian Eno over Salomé Voegelin and Steve Goodman, to Louis Chude-Sokei, Fred Moten, Thomas Meinecke and Erik Steinskog, again sidestepping non-genealogically to Gloria Evangelina Anzaldúa, Karen Barad, Christoph Cox, Jacques Derrida, Drexciya, Karl Marx and Friedrich Engels, to Nick Land, Sun Ra, Klaus Theweleit, Marie Thompson and Alexander Weheliye.

Heuristics of the Sonic

Sonic fiction is all around. It is malleable and plastic, versatile yet demanding, it questions and attacks your authorship and invented heritage or traditions. It expands beyond belief and imagination what you might tell to yourself or others as your personal intellectual or biographical history. Suddenly, you seem to remember and you seem even to sense the effects of events and encounters that might not have been documented on any of the current surveillance files, stored about you.

> So then they called my name, and I realised I was alone, a long way from here, and I don't know what they wanted of me – and I stayed up in the dark. And they called my name again, but I refused to answer. (Sun Ra quoted in Sinker 1992: 30)

The six variations and specimens of sonic fictions enlisted above are not identical in detail. However they are energized by the same drive and by a similar goal: they take the ferment of sonic fiction to infuse their critical or their activist practices – and as a consequence these practices are then accelerated and dynamized in a way that seems to propel them into another modality of critique or activism. This modality is instantaneous and it teleports you into another state of existence, of activities and of connections, apparently.

> And all at once they teleported me down to where they were. In one split second I was up there; next I was down here. So they got that power. (Sun Ra quoted in Sinker 1992: 30)

This disrupting relocation and this dynamization of your existence, thinking, your reflections and actions might then take an effect on your sensing and relating your sensory experiences with your skilled technopractices. Like in the case of Jaki Liebezeit as observed by Holger Czukay and narrated by Kodwo Eshun:

> Czukay said that as Liebezeit's drumming became simpler, he started drumming like the first man who ever drummed, like a stone age man. And the more simple he got, the more he started

to sound like a machine. I was really amazed by this, because it conjured up this image of a drum machine in the Palaeolithic age. Suddenly you start imagining 2001, and instead of this monolith you see this 808 drum machine with no surface, this impalpable surface, landing, and these ape men start touching it. (Eshun quoted in Weelden 1999)

Ape men, already humanoid aliens, touching the monolith, encountering the skilled knowledge of this sublime machine. Could you and I exist as such a soft machine?

Then they talked to me, they had antennas, and they had red eyes that glow like that. And they wanted me to be one of them, and I said no, it's natural for you to be like that, but it might hurt me if you gave me some. (Sun Ra quoted in Sinker 1992: 30)

However, if one would engage in such a sensory practice of listening or sounding, of sonicking or resonating, maybe of drumming – how would that affect my existence and being, my knowing and sensing and thinking? Could the music emerging from this moment represent the sound of an imaginary place that did not yet exist but would soon? Encountering a *sonic thing*? The implex of a sound?

The sonic thing is not through its autonomy but is its action as interaction, creating not itself but the event of the moment, the aesthetic moment of the work and of the everyday as the commingling of what there is apart. (Voegelin 2014: 98)

So through this idea of Jurassic drumming, it suddenly seemed to me that producers had a much clearer idea of the science fiction capacities of their music. Suddenly it was evident that 'sonic fiction', as I proposed it, was already being practised by producers, musicians and composers. All I had to do was extract what was already there and materialise it. (Eshun quoted in Weelden 1999)

The sounds and the thoughts materialize. They coalesce to sonic fiction. They fall in place, apparently:

All the ideas seemed to rush towards this – sonic fiction seemed to be an attractor – and all the terms just moved towards it and it was the easiest thing in the world to extract them and plug them all into each other. (Eshun quoted in Weelden 1999)

Sonic fiction does not write a theory about sound or on sound, but *through* sound. Sonic fiction represents a sort of *sonology*: this is how Kodwo Eshun understands his achievement in writing:

So it becomes a sonology of history, not a historical contextualisation of sound. (Eshun quoted in Weelden 1999)

From the current divergent auralities and their alien sonic nontologies emerges – through some mixillogics of syrrhesis fiction, incorporating the mythscience of ultrablack resistance – at some point, apparently, the subsequent mutantextures of a decolontological rhythmight and its acid communist sonology:

What it [this concept] means to us is not explained at length, but is shown in the scenes mentioned, in the wild and in action. (Dath & Kirchner 2012: 15;[1] translated by Holger Schulze)

This book did not start with an introduction but an extradition. Therefore, it also does not end with a conclusion but an inconclusion (cf. Eshun & Sagar 2007: 15). Think of this book as an inventory of some of the potential thoughts and imaginations connected to the concept of sonic fiction; maybe it is a first registration of its effects, a sort of opening encounter with these six affects, intensities, passions, commitments, risks, gambles and demands of sonic fiction.

Anyway, they talked to me about this planet, and the way it was headed and what was going to happen to teenagers, and governments, and people. They said they wanted me to talk to them. And I said I wasn't interested. (Sun Ra quoted in Sinker 1992: 30)

This is where sonic fiction begins. Right here, right now. Radiating, generating. In your sonics and in your fictions.

NOTES

Extradition

1   Original quote: 'Schält sich diese Stimme noch einmal aus dem langsam verglimmenden Synthesizer-Arpeggio heraus und vollendet den Satz, den sie immer wieder begonnen hatte, nur um von stolpernden Chops des "Think"-Breaks unterbrochen zu werden.'

Chapter 1

1   Original quote: '"Heller als die Sonne" funktioniert merkwürdigerweise tatsächlich eher als langes Review, denn als Theoriebuch, eher als Musterbeispiel der Anwendung diverser Theorien, die auch in der Musik selber im Umlauf sind' (Kösch 1999).
2   Original quote: 'Wo und wann ist im Einzelfall der entscheidende Augenblick, in dem ich die Mittel (mein Wissen und Können, mein Vermögen) nicht mehr beherrsche, sondern ins Spiel bringen und loslassen soll? ... Wie, nach welchen Kriterien führe ich es dann fort? Und wann ist es fertig?'

Chapter 2

1   Original quote: 'Alle Stellen (ich hab sie jetzt nicht im Kopf, es waren aber nicht wenige), die im Duktus und der Wortwahl so ein bisschen an die New-Wave-Science-Fiction der sechziger/siebziger (den "New World" – Sound, Moorcock, Ballard etc.) erinnerten, gingen sehr schnell, das ist der Tonfall, mit dem ich ja selber als Science-Fiction-Leser aufgewachsen war, auch einer bestimmten Tonart der entsprechenden deutschen Übersetzungen.'

2   Original quote: 'Mehrere Sachen im Zusammenhang mit (Free)
     Jazz musste ich mir etwas genauer überlegen; ich wollte nicht blind,
     d.h. nach Vorstellung, wie etwas klingt, wovon K.E. schreibt, diese
     Passagen eindeutschen, und habe daher damals eine Art Crash-Kurs
     in diesen Dingen absolviert, insbesondere Alice Coltrane kannte ich
     praktisch überhaupt nicht, das war ein großer persönlicher Gewinn
     – und von Sun Ra wusste ich nur klischeehaftes Zeug, das sich beim
     genaueren Hören dann in, hoffe ich, etwas besseres Verständnis
     fortentwickeln ließ.'
3   Original quote: 'Die smartesten Produzenten der Neuzeit haben
     das, was die Schweden der Menschheit hinterließen, immer nur als
     Fest des Leichten und Graziösen, als etwas Reines, Heiliges, eine
     erdfern durch das All schwebende fettglänzende Hochzeitsnudel
     gefeiert. Madonna aber legt für "Hung up" die andere, die dreckige
     und fordernde, kurz: die brutale Seite der "Abba"-Erfahrung frei,
     das tonnenschwere Kettenfahrzeug der Liebe, den High-Tech-
     Tanzpanzer.'
4   Original quote: 'Und so geht es weiter, auf durchgängig gehaltenem
     hohen Niveau: "Get Together" klingt wie unter Wasser von
     denkenden Badezusätzen auf Atom-U-Boot-Navigationscomputern
     programmiert, "Sorry" holt uralte Bässe aus dem Keller der
     Pyramiden und beschießt damit die Wolken, "Future Lovers"
     jongliert akustische Magnetfelder und malt die Nacht mit
     Stroboskoplicht an, "I Love New York" baut eine tönende Stadt
     aus rhythmisch sortierten Hitzewallungen zwischen steilen
     Betonwänden – es geht, sagt dies alles, insgesamt um Synästhetisches.
     Bilder und Düfte sind immer mitgedacht.'
5   Original quote: 'Die Kernfrage lautet, ob so etwas wie sozialer
     Fortschritt gedacht und, wichtiger, gemacht werden kann. Man
     könnte sagen, dass das Buch eine Art Roman in Begriffen ist: Es
     begleitet die Schicksale von Versuchen, die Welt besser einzurichten,
     als die neuzeitlichen Menschen sie vorfanden, als sie anfingen,
     neuzeitliche Menschen zu sein.'
6   Original quote: 'Wie in jedem historischen Roman kommt auch hier
     die Liebe vor. Held des Buches ist aber ein Begriff, den wir bei Paul
     Valéry gefunden und dann für andere Zwecke als seine angereichert
     und verändert haben: der Implex. Was er bei uns bedeutet, wird
     nicht langwierig erklärt, sondern auf den genannten Schauplätzen
     gezeigt, in freier Wildbahn und in Aktion.'
7   Original quote: 'Der Kapitalismus ist für Marx das historisch
     einmalige Ereignis einer Form von Ausbeutung, die so viel
     Reichtum produziert, dass die Abschaffung der Aubeutung auf die
     Tagesordnung genommen weren kann. Sieht man das vorhandene
     Falsche nicht einfach als einen Fehler an, der aufgrund falscher Ideen

in die Irre geht, sondern als einziges vorhandenes Reservoir für die richtige Praxis dann wird man sich über Leute, die glauben es würde schon genügen, den Menschen die falschen Ideen auszutreiben, eher lustig machen.'

8   Original quote: 'Daß die Wasch- oder die Geschirspülmaschine der Misogynie ein paar Waffen aus der Hand geschlagen hat, war allerdings nirgends und niemals hinreichend für die entsprechenden gesellschaftlichen Veränderungen; in dieser Detailbeobachtung steckt bereits alles, was man etwa über die Chancen der weiteren Beseitigung arbeitsteiliger Nährböden für Hierarchien, Ausbeutungsverhältnisse, Ausgrenzung und so weiter wissen sollte.'

9   Original quote: 'Es gibt, sagt er, ein Ding nur dann, wenn man etwas damit machen kann, und man hat dann ein richtiges Bild von diesem Ding, wenn man aufgrund dieses Bildes das, was man machen will, auch tatsächlich erfolgreich machen kann.'

10   Original quote: 'Eine Enzyklopädie der historischen Möglichkeiten, realisierter und verpasster; von Befreiungsbewegungen, ihren materiellen Voraussetzungen und den Gründen für ihr Scheitern; ein Kompendium von Theorien, ungenutzten und solchen, deren Gültigkeitsdatum abgelaufen ist. Eine dialektische Lehre des Nachdenkens über den Fortschritt, ein Insistieren auf der Vernunft in der Geschichte – die keine Leiter ist, sondern ein mindestens vierdimensionales ungerichtetes Ensemble von Möglichkeiten und Situationen. Ein Arsenal geschärfter Instrumente der Kritik: Kritik an Ideologien, am bequemen Denken, am Überhauptnicht-Denken.'

11   Original quote: 'Wie kommt uns die Zukunft entgegen und können wir ihr auf halbem Weg begegnen? Wie lassen sich Freiheit und Emanzipation jenseits einer antiquierten Fortschrittslogik denken, und zwar (von) außerhalb Europas oder Nordamerikas – und lässt sich Fortschritt überhaupt noch anders denken, als ausgehend von Kulturen, denen in traditionellen westlichen – und nicht nur in explizit kolonialen und rassistischen – Diskursen die Fähigkeit Geschichte zu produzieren oder zu haben abgesprochen wurde? Und muss nicht, wer heute über politischen Fortschritt oder Emanzipation nachdenkt, das politisches Subjekt des 21. Jahrhunderts miteinbeziehen,den Flüchtling?'

Chapter 3

1   Original quote: 'Afrofuturismus bringt die Idee einer schwarzen Geheimtechnologie in Anschlag, um Momente spekulativer Beschleunigung zuerzeugen. ◊ Blackzelerationismus behauptet,

dass es auf dem Territorium des Schwarzseins schon immer einen Akzelerationismus gegeben hat, bewusst oder nicht. ◊ Sinofuturismus kartographiert die Nachtseite des tumultösen Aufschwungs in Ostasien, indem er heterogene Versatzstücke zu einer Topologie des planetaren Kapitalismus verknotet. ◊ Shanghai-Futurismus wettet letztlich darauf, dass es gelingt, sich von der üblichen Auffassung vom Wesen der Zeit zu lösen. ◊ Golf-Futurismus produziert eine seltsame Mitose, jenseits von Masterplanern und Architekten, während er die Spaltung von Welten in ein vorher und nachher, wir und sie, real und nicht real vorantreibt. ◊ Die globale Dubaifizierung ist schon in vollem Gange, sie legt weiter zu und gibt alles, um ihre Mission mit Lichtgeschwindigkeit zu vollenden.'

Chapter 4

1 Original quote: 'Analog dazu lassen sich Erkenntnisräume wie der wissenschaftliche, der philosophische und der ästhetische vergrößern, indem man jeden davon in anderen nachbaut. Tut man dies beim Schreiben, dann muss man sowohl Abhandlungen wie Erzählungen, Gedichte wie Manifeste, Analysen wie Spekulationen verfassen – und zwar Gedichte über Analysen, Spekulationen über Erzählungen und so weiter.'

Chapter 5

1 Original quote: 'In der Geschichtsschreibung sind Wahrheit und mit Affekt besetzte Fiktion ebenso schwer (und manchmal gar nicht) zu unterscheiden … GHOSTS: – das war vor 30 Jahren ein Stück aus dem Tenorsaxophon Albert Aylers … *höchst wirklich* … heute etwas, worauf Michael jackson tanzt … *they've come a long long way*.'

Chapter 6

1 Original quote: 'Wir haben hier die Möglichkeit sozialistisch zu quatschen. Einige können evolutionär reden, andere dürfen revolutionär reden, ja. Und was passiert objektiv? An der Unterdrückung ändert sich überhaupt nichts! Fernsehen ist ein

Unterdrückungsinstrument in dieser Massengesellschaft! Und deswegen ist es ganz klar hier, wenn überhaupt noch was passieren soll hier, muss man sich gegen den Unterdrücker stellen. Man muss parteiisch sein. Das muss man hier einfach sagen. Und deswegen mach ich jetzt hier diesen Tisch mal kaputt. Ja, damit man mal genau Bescheid weiß!'

2   Original quote: 'Sie ist Katastrophenwissenschaft als ein Akt, der die formalen Strukturen von Raum und Zeit zerlegt. In der Mimikry dieser Wissenschaft an die elektronische Musik kollabieren sowohl in der Wissenschaft als auch in der Musik die formalen Strukturen der Zeit, regredieren zu Schlamm, und der Raum wird hin und hergeschoben, bis er sich krümmt, um von den Pulsationen der Alien-Musik zertrampelt zu werden, während der Kopfraum seekrank wird.'

3   Original quote: 'Der entscheidende Schritt ist hier die Konstruktion des exklusiven Gegenteils. Underground Resistance sagen irgendwo, Verschwinden sei unsere Zukunft, und nach Eshun sollte damit die Black Power von UR unsichtbar sein, nicht identifizierbar, verborgen, unkenntlich und nicht öffentlich.'

4   Original quote: 'Damit ist die Kriegsmaschine umfassend in Anschlag gebracht. Die Guerilla kämpft als eine Meute von Maschinisten mit technischen Apparaten gegen den maschinell urbanen Maschinenkörper des Kapitals. Auch Schrift und Musik können Kriegsmaschinen sein.'

5   Original quote: 'Sonisches Denken oder Non-Musikologie komponiert die Theorie als ihr eigenes Objekt, schreibt eine autonome Musik-Fiktion.... Fiktion impliziert Performance, Erfindung, Artefakt und Konstruktion, aber dies in einem nicht-expressiven und nicht-repräsentationalen Sinn, sondern als Immanenz.'

6   Original quote: 'Nicht-Musikologie fordert keineswegs eine neue Musikologie, sondern eine generische Wissenschaft der Musik, oder, um es anders zu sagen, keine Wissenschaft, sondern eher eine Häresie oder eine Fiktion im Angesicht der Musik.'

7   Original quote: 'Radikale Musik gleicht einer Art von Blackbox: sie ist eine Musikbox der und für die Blackness, und der Theoretiker und der Konsument der Musik nehmen selbst einen Platz in der Blackbox ein und treten nicht von außen an die Box heran. Es gibt eine nicht-musikalische Triangularität zu vermelden: Der (multiple) Produzent, der die Transversalität des Schwarzen zum Klingen bringt; die schwarze Musikbox als ein unendliches Klingen des Nichtfassbaren/Schwarzen; der Konsument, der Teile aus dem Unendlichen der schwarzen Musikbox heraus hört.'

Inconclusion

1 Original quote: 'Was er [dieser Begriff] bei uns bedeutet, wird nicht langwierig erklärt, sondern auf den genannten Schauplätzen gezeigt, in freier Wildbahn und in Aktion.'

REFERENCES

Adams, D. (1979), *The Hitchhiker's Guide to the Galaxy*, London: Pan Books.

Ahmed, S. (2007), 'A Phenomenology of Whiteness', *Feminist Theory* 8 (2): 149–168.

Anzaldúa, G. E. (1983), 'La prieta', in C. Moraga & G. Anzaldúa (eds), *This Bridge Called My Back*, New York: Kitchen Table/Women of Color Press, 198–209.

Anzaldúa, G. E. [1986] (2009), 'Creativity and Switching Modes of Consciousness', in AnaLouise Keating (ed.), *The Gloria Anzaldúa Reader*, Durham NC: Duke University Press, 103–104.

Anzaldúa, G. E. [1987] (2012), *Borderlands/La Frontera: The New Mestiza/4th edition*, San Francisco: Aunt Lute Books.

Anzaldúa, G. E. (1990), *Making Face, Making Soul/Haciendo Caras: Creative and Critical Perspectives by Feminists of Color*, San Francisco: Aunt Lute Books.

Anzaldúa, G. E. (1999), 'Putting Coyolxauhqui Together: A Creative Process', in Marla Morris, Mary Aswell Doll & William F. Pinar (eds), *How We Work*, New York: Peter Lang, 242–261.

Anzaldúa, G. E. (2000), 'Doing Gigs', in A. Keating (ed.), *Gloria Anzaldúa: Interviews/Entrevistas*, New York: Routledge Publishing, 211–234.

Anzaldúa, G. E. (2002), 'Now Let Us Shift … the Path of Conocimiento … Inner Works, Public Acts', in G. Anzaldúa & A. Keating (eds), *This Bridge We Call Home: Radical Visions for Transformation*, New York: Routledge Publishing, 540–578.

Auinger, S. and Odland, B. (2007), 'Hearing Perspective (Think with Your Ears)', in *Sam Auinger & Friends. A Hearing Perspective. Die Sound- und Videoinstallationen des Sam Auinger. Edited by Carsten Seiffarth and Martin Sturm*, OK Linz / Folio Verlag, Wien, 2008, Vienna: Folio.

Avanessian, A. & Moalemi, M., eds (2018), *Ethnofuturismen*, Berlin: Merve Verlag.

Bachelard, G. [1950] (2000), *The Dialectic of Duration*, Manchester: Clinamen.

Badiou, A. (2016), 'Reflections on the Recent Election', Verso, 15 November 2016 – online: https://www.versobooks.com/blogs/2940-

alain-badiou-reflections-on-the-recent-election [accessed 21 December 2018].

Badley, L. (1995), *Film, Horror and the Body Fanstastic*, Westport, CT: Greenwood Press.

Barad, K. (2003), 'Posthumanist Performativity: Toward an Understanding of How Matter Comes to Matter', *Signs* 28 (3): 801–831.

Barad, K. (2007), *Meeting the Universe Halfway: Quantum Physics and the Entanglement of Matter and Meaning*, Durham NC: Duke University Press.

Barad, K. (2011), 'Nature's Queer Performativity', *Qui Parle: Critical Humanities and Social Sciences* 19 (2): 121–158.

Barad, K. (2012), 'Interview', in Rick Dolphijn & Iris Van der Tuin (eds), *New Materialism: Interviews & Cartographies*, Ann Arbor: Open Humanities Press, 48–70.

Baudrillard, J. [1981] (1994), *Simulacra and Simulation*, Ann Arbor: University of Michigan Press.

Bauer, G. & Stockhammer, R., eds (2000), *Möglichkeitssinn. Phantasie und Phantastik in der Erzählliteratur des 20. Jahrhunderts*, Opladen: Westdeutscher Verlag.

Boards of Canada (1998), *Music Has the Right to Children*, Sheffield: WARP Records.

Burghart, C. (2013), *Paul Valérys Blick auf den modernen Menschen: Experiment einer neuen Philosophie*, Berlin: Frank & Timme.

Burial (2006), *Untrue*, London: Hyperdub.

Burroughs, W. S. (1961), *The Soft Machine*, Paris: Olympia Press.

Burroughs, W. S. (1962), *The Ticket That Exploded*, Paris: Olympia Press.

Butler, O. (1997), *Dawn (Xenogenesis, Bk. 1)*, New York: Warner Aspect.

Calderón, D., Delgado Bernal, D., Pérez Huber, L., Malagón, M.C. & Vélez, V. N. (2012), 'A Chicana Feminist Epistemology Revisited', *Harvard Educational Review* 82(4): 513–539.

Campt, T. M. (2014), 'Black Feminist Futures and the Practice of Fugitivity', *Helen Pond McIntyre '48 Lecture*, Barnard College, 7 October, online at: http://bcrw.barnard.edu/videos/tina-campt-black-feminist-futures-and-the-practice-of-fugitivity/ [accessed 21 December 2018].

Campt, T. M. (2017), *Listening to Images*, Durham, NC: Duke University Press.

Carstens, D. & Land, N. (2009): 'Hyperstition: An Introduction', *Merliquify*, online at: http://merliquify.com/blog/articles/hyperstition-an-introduction/#.W365rM4zZEY [accessed 21 December 2018].

Chew-Bose, D. (2017), *Too Much and Not the Mood: Essays*, New York: Farrar, Straus and Giroux.

Chude-Sokei, L. (2016), *The Sound of Culture: Diaspora and Black Technopoetics*, Middletown, CT: Wesleyan University Press.

Clarke, A. C. (1962), *Profiles of the Future: An Inquiry Into the Limits of the Possible*, Boston: Houghton Mifflin Harcourt Publishing.

Cobussen, M., Schulze, H. & Meelberg, V. (2013), 'Editorial', in M. Cobussen, V. Meelberg & H. Schulze (eds), 'Towards New Sonic Epistemologies', Special Issue of *Journal of Sonic Studies* 3(4) online: https://www.researchcatalogue.net/view/266013/266014.

Coleman, J. (1996), *Public Reading and the Reading Public in Late Medieval England and France*, Cambridge: Cambridge University Press.

Coleman, J. (2007), 'Aurality', in P. Strohm (ed.), *Middle English*, Oxford: Oxford University Press, 68–85.

Collins, L. (1972), *Think (About It)*, London: People.

Colquhoun, M. (2018), 'Acid Communism', in 'Marx from the Margins: A Collective Project, from A to Z', Special Issue of *Krisis – Journal for Contemporary Philosophy*, (2), online at: https://web.archive. org/web/20181116222624/http://krisis.eu/acid-communism/ [accessed 21 December 2018].

Cox, C. (2011), 'Beyond Representation and Signification: Toward a Sonic Materialism', *Journal of Visual Culture* 10 (2): 145–161.

Cox, C. (2013), 'Sonic Philosophy', *ArtPulse* 4 (16), online at: http:// artpulsemagazine.com/sonic-philosophy [accessed 21 December 2018].

Cox, C. (2014), 'Afrofuturism, Afro-Pessimism and the Politics of Abstraction: An Interview with Kodwo Eshun' http://faculty. hampshire.edu/ccox/Cox.Interview%20with%20Kodwo%20Eshun. pdf.

Cox, C. (2018), *Sonic Flux: Sound, Art, and Metaphysics*, Chicago: University of Chicago Press.

Dahms, E. A. (2012), *The Life and Work of Gloria Anzaldua: An Intellectual Biography*, Lexington: University of Kentucky Theses and Dissertations–Hispanic Studies. 6, online at: https://uknowledge.uky. edu/hisp_etds/6 [accessed 21 December 2018].

Dath, D. (1995), *Cordula killt Dich! oder Wir sind doch nicht Nemesis von jedem Pfeifenheini. Roman der Auferstehung*, Berlin: Verbrecher Verlag.

Dath, D. (2001), *Phonon oder Staat ohne Namen. Roman*, Berlin: Verbrecher Verlag.

Dath, D. (2005a), *Die salzweißen Augen. Vierzehn Briefe über Drastik und Deutlichkeit*, Frankfurt am Main: Suhrkamp Verlag.

Dath, D. (2005b), *Für immer in Honig. Roman*, illustriert von Daniela Burger, Implex, Berlin (erste Fassung, 971 Seiten), 2, überarbeitete Auflage, Verbrecher Verlag, Berlin 2008, (1035 Seiten).

Dath, D. (2005c), 'Sie malt die Nacht mit Licht an', *Frankfurter Allgemeine Zeitung* 57(264): 12, 33.

Dath, D. (2007), *Heute keine Konferenz. Texte für die Zeitung*, Frankfurt am Main: Suhrkamp Verlag.

Dath, D. (2008), *Die Abschaffung der Arten*. *Roman*, Frankfurt am Main: Suhrkamp Verlag.

Dath, D. (2014), *Feldeváye*. *Roman*, Berlin: Suhrkamp Verlag.

Dath, D. (2015), *Venus siegt*. *Roman*, Frankfurt am Main: Fischer Tor.

Dath, D. (2018a), *Karl Marx*, Stuttgart: Philipp Reclam jun.

Dath, D. (2018b), *Re: Nachfrage: Übersetzung von 'More Brilliant than the Sun'*, 25 October (personal email).

Dath. D. & Greffrath, M. (2018), *Das Menschen Mögliche. Zur Aktualität von Günther Anders – Verleihung des Günther Anders-Preises an Dietmar Dath am 12. März 2018*, Vienna: Picus Verlag.

Dath, D. & Kirchner, B. (2012), *Der Implex. Sozialer Fortschritt: Geschichte und Idee*, Frankfurt am Main: Suhrkamp Verlag.

Davis, C. (2005), 'Hauntology, Spectres and Phantoms', *French Studies* 59(3): 373–379.

Dehghani, M., Kamrani, E., Salarpouri, A. & Sharifian, S. (2016), 'Otolith Dimensions (Length, Width), Otolith Weight and Fish Length of Sardinella Sindensis (Day, 1878), as Index for Environmental Studies, Persian Gulf, Iran', *Marine Biodiversity Records* 9: article 44. https://doi.org/10.1186/s41200-016-0039-0.

Delany, S. (1984), *Stars in My Pocket Like Grains of Sand*, New York: Bantam Books.

Derrida, J. (1972), 'Les sources de Valéry. Qual, Quelle', *MLN - Modern Language Notes*, Vol. 87, No. 4, French Issue: Paul Valery (May, 1972), pp. 563–599.

Derrida, J. (1994), *Specters of Marx: The State of the Debt, the Work of Mourning and the New International*, London: Routledge Publishing.

Dery, M. (1994), 'Black to the Future: Interviews with Samuel R. Delany, Greg Tate, and Tricia Ros', in M. Dery (ed.), *Flame Wars: The Discourse of Cyberculture*, Durham NC: Duke University Press. 179–222.

Dery, M. (2016), 'Black to the Future: Afrofuturism 1.0', *Fabrikzeitung*, Zürich: IG Rote Fabrik, online at: https://www.fabrikzeitung.ch/black-to-the-future-afro-futurism-1-0/#/ [accessed 21 December 2018].

Diederichsen, D., ed. (1998), *Loving the Alien: Science Fiction, Diaspora, Multikultur*, Berlin: ID-Verlag.

Dillon, B. (2017), *Essayism: On Form, Feeling, and Nonfiction*, London: Fitzcarraldo.

Dix, H., ed. (2018), *Autofiction in English*, London: Palgrave Macmillan.

Drexciya (1997), *The Quest*, Detroit, MA: Submerge (3).

Eagleton, T. (2016), *Materialism*, Yale: Yale University Press.

Ellison, R. (1995), *The Collected Essays of Ralph Ellison*, ed. J. F. Callahan, New York: Modern Library.

Erlmann, V. (2010), *Reason and Resonance. A History of Modern Aurality*, New York: ZONE Books.

Eshun, E. (2005), Black Gold of the Sun: Searching for Home in England and Africa, London: Hamish Hamilton.

Eshun, K. (1992a), 'Club Licks: Kodwo Eshun hits the floor', *The Wire* 105 (November): 69.

Eshun, K. (1992b), 'Club Licks: Kodwo Eshun Keeps it Street (Oh, Shut Up - Ed)', *The Wire* 104 (October): 68.

Eshun, K. (1992c), 'Club Licks: Kodwo Eshun Trips the Light Fantastic', *The Wire* 101 (July): 66–67.

Eshun, K. (1993a), 'Club Trax: Kodwo Eshun Does a Critical Beatdown on the New Dance Releases', *The Wire* 116 (October): 69.

Eshun, K. (1993b), 'Club Trax: Kodwo Eshun Finds Hidden Depths in New Techno, HipHop and Streetsoul Melodies', *The Wire* 112 (June), 71.

Eshun, K. (1995), 'Club Trax: Kodwo Eshun Speeds through April's Hottest Wax', *The Wire* 134 (April): 64.

Eshun, K. (1998), *More Brilliant than the Sun: Adventures in Sonic Fiction*, London: Quartet Books.

Eshun, K. (1999), *Heller als die Sonne. Abenteuer in der Sonic Fiction*. Übersetzt von Dietmar Dath, Berlin: ID-Verlag.

Eshun, K. (2003), 'Further Considerations on Afrofuturism', *CR: The New Centennial Review* 3 (2): 287–302, online at: https://www.kit.ntnu.no/sites/www.kit.ntnu.no/files/KodwoEshun_Afrofuturism_0.pdf [accessed 21 December 2018].

Eshun, K. (2012), *Dan Graham: Rock My Religion*, Cambridge, MA: MIT-Press.

Eshun, K. (2018a), 'Mark Fisher Memorial Lecture', Goldsmiths, University of London, 19 January 2018, online at: https://www.youtube.com/watch?v=ufznupiVCLs [accessed 21 December 2018].

Eshun, K. (2018b), *More Brilliant than the Sun: Adventures in Sonic Fiction*, Introduction by S. Goodman, New York: Verso Books.

Eshun, K. & Pomassl, F. (1999), *Architectronics*, Vienna: Craft Records.

Eshun, K. & Sagar, A., eds. (2007), *The Ghosts of Songs: The Film Art of the Black Audio Film Collective 1982–1998*, Liverpool: Liverpool University Press.

Eshun, K. & Sagar, A. (2009), *A Long Time Between Suns*, ed. A. Colin & E. Pethick, Contributions by D. McCarty, J. Matthee & T.J. Demos, Berlin: Sternberg Press.

Eshun, K. & Sagar, A. (2015), *World 3*, Bergen: Yydap Pub.

Fisher, M. (2000), *Flatline Constructs: Gothic Materialism and Cybernetic Theory-Fiction*. University of Warwick. PhD Thesis. https://web.archive.org/web/20101123101648/http://cinestatic.com/trans-mat/Fisher/FCcontents.htm (08/08/2016).

Fisher, M. (2005), 'Simon's Interview with CCRU (1998)', *k-punk*, 20 January 2005, online at: http://k-punk.abstractdynamics. org/archives/004807.html [accessed 21 December 2018].

Fisher, M. (2009), *Capitalist Realism: Is There No Alternative?*, Winchester: Zero Books.

Fisher, M. (2011), 'Nick Land: Mind Games', *Dazed*, 1 June 2011, online at: http://www.dazeddigital.com/artsandculture/article/10459/1/nick-land-mind-games [accessed 21 December 2018].

Fisher, M. (2012), 'What Is Hauntology?', *Film Quarterly* 66 (1) (Fall): 16–24.

Fisher, M. (2013), 'The Metaphysics of Crackle: Afrofuturism and Hauntology', *Dancecult: Journal of Electronic Dance Music Culture* 5 (2): 42–55.

Fisher, M. (2014a), *Ghosts of My Life: Writings on Depression, Hauntology and Lost Futures*, Winchester: Zero Books.

Fisher, M. (2014b), 'Terminator vs. Avatar: Notes on Accelerationism (presented at the Accelerationism symposium, Goldsmiths: 14:09:2010)', online at: http://markfisherreblog.tumblr. com/post/32522465887/terminator-vs-avatar-notes-on-accelerationism [accessed 21 December 2018].

Fisher, M. (2017), *The Weird and the Eerie*, London: Repeater Books.

Fisher, M. (2018), *k-punk: The Collected and Unpublished Writings of Mark Fisher (2004–2016)*, ed. D. Ambrose, foreword by S. Reynolds, London: Repeater Books.

Fowler, J. (2015), 'JMF075', online: http://jarrodfowler.com/JMF075.html [accessed 21 December 2018]

Galloway, A. (2014), *Laruelle: Against the Digital*, Minneapolis: University of Minnesota Press.

Gasparini, P. (2008), *Autofiction: Une aventure du langage*, Paris: Editions de Seuil.

Gautier, A. M. O. (2014), Aurality: *Listening and Knowledge in Nineteenth-Century Colombia*, Durham, NC: Duke University Press.

Geerts, E. and Tuin I.V.D. (2016), 'Diffraction & Reading Diffractively: 27. Juli 2016', in David Gauthier and Sam Skinner (eds), *New Materialism: How Matter Comes To Matter – Almanac*, online at: https://web.archive.org/web/20180829224951/http://newmaterialism. eu/almanac/d/diffraction [accessed 21 December 2018]

Gilroy, P. (1993), *The Black Atlantic: Modernity and Double Consciousness*, Cambridge, MA: Harvard University Press.

Goh, A. (2017), 'Sounding Situated Knowledges: Echo in Archaeoacoustics', J. Lavender (ed.), 'Sounding/Thinking', Special Issue of *Parallax* 23 (3): 283–304.

Goodman, S. (2010), *Sonic Warfare: Sound, Affect, and the Ecology of Fear*, Cambridge, MA: MIT Press.

Goodman, S. and Heys, T. (2014), *Martial Hauntology*, London: AUDINT Records.

Grell, J. (2014), *L'Autofiction*, Paris: Armand Colin.

Groth, S. K. and Samson, K. (2016), 'Audio Paper: A Manifesto', 'Fluid Sounds', Special Issue of *Seismograf/DMT*, Copenhagen 30 August 2016, online at: http://seismograf.org/fokus/fluid-sounds/audio_paper_manifesto [accessed 21 December 2018].

Gunkell, H., Hameed, A. & O'Sullivan, S., eds (2017), *Futures and Fictions: Essays and Conversations that Explore Alternative Narratives and Image Worlds that Might Be Pitched Against the Impasses of Our Neo-Liberal Present*, London: Repeater Books.

Gutmair, U. (2000), 'Deprogram! Reprogram! Kalt glitzernde Momente in Kodwo Eshuns Futurhythmachine', *De:Bug*, 1 January 2000.

Hägglund, M. (2008), *Radical Atheism: Derrida and the Time of Life*, Stanford, CA: Stanford University Press.

Hameed, A. (2016), 'Black Atlantis', Goldsmiths, University of London, 17 October 2016, online at: https://www.youtube.com/watch?v=2TcxA04WPzA [accessed 21 December 2018].

Handke, P. (1989), *Versuch Über die Müdigkeit*, Frankfurt am Main: Suhrkamp Verlag.

Harney, S. & Moten, F. (2013), *The Undercommons: Fugitive Planning & Black Study*, Wivenhoe: Minor Compositions.

Havis, D. N. (2009), 'Blackness Beyond Witness: Black Vernacular Phenomena and Auditory Identity', *Philosophy & Social Criticism* 35 (7): 747–759.

Herzogenrath, B., ed. (2017), *Sonic Thinking: A Media Philosophical Approach*, vol. 4 of the book series *Thinking Media*, New York: Bloomsbury Publishing.

Highstein, S. M., Fay, R.R. & Popper, A.N., eds (2004), *The Vestibular System*, Berlin: Springer.

Holt, M. (2017a), 'A Sonic Fiction of Boring Dystopia', PhD diss., Goldsmiths College London.

Holt, M. (2017b), 'The Terrifying Ambivalence of Theory-Fiction', *Ark*, 16 May 2017, online at: http://arkbooks.dk/the-terrifying-ambivalence-of-theory-fiction/ [accessed 21 December 2018].

Holt, M. (2019), *Pop Music and Hip Ennui: A Sonic Fiction of Capitalist Realism*, New York: Bloomsbury Publishing.

Husserl, E. [1913/21] (1984), *Logische Untersuchungen, Zweiter. Band, Erster Teil: Untersuchungen zur Phänomenologie und Theorie der Erkenntnis – Herausgegeben von Ursula Panzer. Husserliana: Gesammelte Werke XIX/1*, Den Haag: Martinus Nijhoff Publishers.

Ikoniadou, E. (2016), 'The Unsound Methods of AUDINT', *The Wire*, December, online at: https://www.thewire.co.uk/in-writing/interviews/the-unsound-methods-of-audint [accessed 21 December 2018].

Jasen, P. C. (2016), *Low End Theory: Bass, Bodies and the Materiality of Sonic Experience*, New York: Bloomsbury Press.

Kapchan, D., ed. (2017), *Theorizing Sound Writing*, Middletown CT: Wesleyan University Press.

Kiberd, R. (2015), 'The Rise and Fall of "Boring Dystopia," the Anti-Facebook Facebook Group', *Motherboard* – online: http://motherboard.vice.com/read/the-rise-and-fall-of-boring-dystopia-the-anti-facebook-facebook-group. [accessed 21 December 2018].

Kittler, F. (1985), *Aufschreibesysteme 1800–1900*, Munich: Wilhelm Fink.

Kniep R., Zahn D., Wulfes J. & Walther, L.E. (2017), 'The Sense of Balance in Humans: Structural Features of Otoconia and their Response to Linear Acceleration', *PLoS ONE* 12 (4): e0175769. https://doi.org/10.1371/journal.pone.0175769.

Knoblauch, J. (2017), 'Loving the Buchproduktion: Interview zum zehnjährigen Jubiläum des ID-Verlags – Ein Gespräch mit Andreas Fanizadeh', *Contraste* Nr. 174, online at: https://www.idverlag.com/artikel.php?artikelID=7 [accessed 21 December 2018].

Koppe, F. (1977), Sprache und Bedürfnis. Zur sprachphilosophischen Grundlage der Geisteswissenschaften, Stuttgart-Bad Cannstatt: Fromman-Holzboog.

Kösch, S. (1999), 'Kodwo Eshun – Heller als die Sonne', *De:Bug* 4 (23), June, 30.

Kraus, C. (2017), 'Sex, Tattle and Soul: How Kathy Acker Shocked and Seduced the Literary World', *The Guardian*, 8 August 2017, online at: https://www.theguardian.com/books/2017/aug/19/sex-tattle-and-soul-how-kathy-acker-shocked-and-seduced-the-literary-world [accessed 21 December 2018].

LaBelle, B. (2018), *Sonic Agency: Sound and Emergent Forms of Resistance*, London: Goldsmiths; Cambridge, MA: MIT Press.

Laboria Cuboniks (2018), *The Xenofeminist Manifesto. A Politics for Alienation*, New York: Verso Books.

Land, N. (1992), *Thirst For Annihilation: George Bataille and Virulent Nihilism*, London: Routledge.

Land, N. (2011), *Fanged Noumena: Collected Writings 1987–2007*, Introduction by R. Brassier & R. Mackay, Falmouth: Urbanomic.

Land, N. (2013), 'Dark Enlightenment', 10 February 2013, online at: http://www.thedarkenlightenment.com/the-dark-enlightenment-by-nick-land/ [accessed 21 December 2018].

Langguth, J. (2010), 'Proposing an Alter-Destiny: Science Fiction in the Art and Music of Sun Ra', in M.J. Bartkowiak (ed.), *Sounds of the Future: Essays on Music in Science Fiction Film*, Jefferson, NC: McFarland, 148–162.

Laruelle, F. (1988), 'Du noir univers: dans les fondations humaines de la couleur', *La Décision philosophique* 5 (April): 107–112.

Laruelle, F. (1991), 'On the Black Universe: In the Human Foundations of Color', in M. Breu (ed.), *Hyun Soo Choi: Seven Large-Scale Paintings*, New York: Thread Waxing Space, 2–4.

Laruelle, F. (2004), 'A New Presentation of Non-Philosophy', 2 November 2004, online at: https://www.onphi.net/corpus/32/a-new-presentation-of-non-philosophy [accessed 21 December 2018].

The Last Angel of History (1996), [Film] Dir. J. Akomfrah, New York: First Run/Icarus Films.

Lavender, J. (2017), 'Sounding/Thinking', 'Sonic Thinking', Special Issue of *Parallax* 23(3): 245–251.

Lefebvre, H. (2013), *Rhythmanalysis: Space, Time and Everyday Life*, London: Bloomsbury.

Lehnerer, T. (1994), *Methode der Kunst*, Würzburg: Königshausen & Neumann.

Lem, S. [1973] (1985), *Imaginary Magnitude*, Boston: Mariner Books.

Lem, S. [1979] (1999), *A Perfect Vacuum*, translated by M. Kandel, Evanston, IL: Northwestern University Press.

Lewis, G. E. (1996), 'Improvised Music after 1950: Afrological and Eurological Perspectives', *Black Music Research* 16 (1): 91–122.

Lockhart, T. (2006), 'Writing the Self: Gloria Anzaldúa, Textual Form, and Feminist Epistemology', *Michigan Feminist Studies* 20.

Lorde, A. (1984), *Sister Outsider: Essays and Speeches*, Berkeley, CA: Ten Speed Press.

Mackay, R. (2013), 'Nick Land – An Experiment in Inhumanism', *Divus*, 27 February 2013, online at: http://divus.cc/london/en/article/nick-land-ein-experiment-im-inhumanismus [accessed 21 December 2018].

Marcuse, H. [1955] (1998), *Eros and Civilisation: A Philosophical Inquiry into Freud*, Oxford: Routledge.

Márquez, C. M. (1991), 'The Sense of Possibility: On the Ontologico-Eidetic Relevance of the Character (the Experimental Ego) in Literary Experience', in A.-T. Tymieniecka (ed.), Book 4, *New Queries in Aesthetics and Metaphysics: Time, Historicity, Art, Culture, Metaphysics, the Transnatural*, Phenomenology in the World Fifty Years after the Death of Edmund Husserl, Dordrecht: Kluwer Academic, 329–342.

Marx, K. & Engels, F. (1848a), *Manifest der kommunistischen Partei*, London: Office der Bildungs-Gesellschaft für Arbeiter.

Marx, K. & Engels, F. (1848b), *The Communist Manifesto*, translated by S. Moore, Marx/Engels Selected Works, Vol. 1, Moscow: Progress Publishers, 1969, 98–137

Meinecke, T. (2001), *Hellblau*, Frankfurt am Main: Suhrkamp Verlag.

Meinecke, T. (2012), *Pale Blue*, translated by D Bowles, Amazon Crossing.

Mignolo, W. D. (2011), *The Darker Side of Western Modernity: Global Futures, Decolonial Options*, Durham, NC: Duke University Press.

Moten, F. (2003), *In the Break: The Aesthetics of Black Radical Tradition*, Minneapolis: University of Minnesota Press.

Musil, R. [1930] (1978), *Der Mann ohne Eigenschaften*, Reinbek: Rowohlt Verlag.

Nelson, A. (2002), 'Introduction: Future Texts', *Social Text* 71, 20 (2), Summer: 97–113.

Nelson, M. (2009), *Bluets*, Seattle, WA: Wave Books.

Nyong'o, T. (2014), 'Afro-philo-sonic Fictions: Black Sound Studies after the Millennium', *Small Axe* 18 (2): 173–179.

Oliveira, P. (2017), 'Decolonizing the Earview of Design: Listening Anxieties and the Apparatus of Auditory Governance', PhD diss., University of the Arts Berlin.

Origin Unknown (1993), 'Valley of the Shadows', Hornchurch: RAM Records.

The Otolith Group (2018), Website – online: http://web.archive. org/web/20181128230215/http://otolithgroup.org/index. php?m=information

Papenburg, J.G. & Schulze, H., eds. (2016), *Sound as Popular Culture: A Research Companion*, Cambridge, MA: MIT-Press.

Parliament (1975), *Mothership Connection*, Wilmington, DE: Casablanca Records.

Pelleter, M. (2018), 'Futurhythmaschinen. Drum-Machines und die klanglichen Zukünfte auditiver Kulturen im 20. Jahrhundert', PhD diss., Leuphana Universität Lüneburg.

Pitts, A. J. (2016), 'Gloria E. Anzaldúa's Autohistoria-teoría as an Epistemology of Self-Knowledge/Ignorance', *Hypatia* 31. https://doi. org/ 10.1111/hypa.12235.

Plant, S. (1998), *Zeroes and Ones: Digital Women and the New Technoculture*, London: HarperCollins.

Reed, I. (1972), *Mumbo Jumbo*, New York: Doubleday.

Reeves-Evison, T., & Shaw, J.K., eds (2017), *Fiction as Method*, Berlin: Sternberg Press.

Reynolds, S. (2009), 'Energy Flash: Cold Getting Nuum', *Energy Flash Simon Reynolds*, 3 November 2009, online at: http:// energyflashbysimonreynolds.blogspot.de/2009/11/renegade-academia-cybernetic-culture.html [accessed 21 December 2018].

Reynolds, S. (2011), *Retromania: Pop Culture's Addiction to its Own Past*, New York: Faber & Faber.

Schrimshaw, W. (2015), 'Exit Immersion', *Sound Studies* 1 (1): 155–170.

Schulze, H. (2000), *Das aleatorische Spiel. Erkundung und Anwendung der nichtintentionalen Werkgenese im 20. Jahrhundert – Theorie der Werkgenese*, vol. 1, Munich: Wilhelm Fink Verlag.

Schulze, H. (2005), *Heuristik. Theorie der intentionalen Werkgenese – Theorie der Werkgenese*, vol. 2, Bielefeld: transcript Verlag.

Schulze, H. (2006), 'Sprechen über Klang. Vorüberlegungen zu einer künftigen Klanganthropologie', in C. Brüstle, M. Rebstock & G. Klein (eds), *Reflexzonen\Migration. Musik im Dialog VI – Jahrbuch der berliner gesellschaft für neue musik 2003/2004*, Saarbrücken: PFAU-Verlag, 70–76.

Schulze, H. (2007), 'Die Audiopietisten. Eine Polemik', *Kultur und Gespenster* 2: 122–129.

Schulze, H. (2008), 'Hypercorporealismus: Eine Wissenschaftsgeschichte des körperlichen Schalls', P. Wicke (ed.), 'Das Sonische – Sounds zwischen Akustik und .sthetik', Special Issue of *Popscriptum* 16, H. 10, online at: http://www2.hu-berlin.de/fpm/popscrip/themen/pst10/pst10_schulze.pdf [accessed 21 December 2018].

Schulze, H. (2013), 'Adventures in Sonic Fiction: A Heuristic for Sound Studies', M. Cobussen, V. Meelberg & H. Schulze (eds), 'Towards New Sonic Epistemologies', Special Issue of *Journal of Sonic Studies* 3(4).

Schulze, H. (2016), 'Idiosyncrasy as Method: Reflections on the Epistemic Continuum', 'Fluid Sounds', Special Issue of *Seismograf/DMT*, 30 August 2016, online at: http://seismograf.org/fokus/fluid-sounds/idiosyncracy-as-method [accessed 21 December 2018].

Schulze, H. (2017), 'How To Think Sonically? On the Generativity of the Flesh', *Sonic Thinking: A Media Philosophical Approach*, vol. 4 of the book series *Thinking Media*, ed. Bernd Herzogenrath, New York: Bloomsbury, 217–242.

Schulze, H. (2018), *The Sonic Persona. An Anthropology of Sound*, New York: Bloomsbury.

Schulze, H. (2019a), 'Sonic Writing', in M. Bull & M. Cobussen (eds), *The Bloomsbury Handbook of Sonic Methodologies*, New York: Bloomsbury.

Schulze, H. (2019b), *Sound Works: A Cultural Theory of Sound Design*, New York: Bloomsbury.

The Schwarzenbach (2012), *Farnschiffe*, Hamburg: Zickzack.

The Schwarzenbach (2015), *Nicht sterben: Aufpassen*, Perpignan: Staubgold.

Serres, M. (1985), *Philosophie des corps mêlés: Les Cinq Sens*, Paris: Éditions Gallimard.

Serres, M. (2008), *The Five Senses: A Philosophy of Mingled Bodies*, translated by M. Sankey & P. Cowley, New York: Continuum Publishers.

Serres, M. (2014), *Thumbelina: The Culture and Technology of Millennials*, translated by D.W. Smith, Lanham, MD: Rowman & Littlefield International.

Sinker, M. (1992), 'Loving the Alien in Advance of the Landing: Black Science Fiction', *The Wire*, 96 (February): 30–33.

Stadler, G. (2015), 'On Whiteness and Sound Studies', *Sounding Out!*, 6 July 2015, online at: https://soundstudiesblog.com/2015/07/06/on-whiteness-and-sound-studies/ [accessed 21 December 2018].

Stanitzek, G. (2011), *Essay: BRD*, Berlin: Vorwerk 8.

Steinbach, A. & Szepanski A. (2017), *Ultrablack of Music: Feindliche Übernahme*, Leipzig: Spector Books.

Steinskog, E. (2018), *Afrofuturism and Black Sound Studies: Culture, Technology, and Things to Come*, Basingstoke: Palgrave Macmillan.

Sterne, J. (2012), *The Sound Studies Reader*, London: Routledge.

Stoever, J. L. (2016), *The Sonic Color Line: Race and the Cultural Politics of Listening*, New York: New York University Press.

Sun Ra [1972] (2010), *The Outer Darkness, Series: Space Poetry – Volume Three*, New York: Norton Records.

Sun Ra (2005), *The Immeasurable Equation: The Collected Poetry and Prose*, ed. J. L. Wolf & H. Geerken, *Introductions and Essays* by J. L. Wolf, H. Geerken, S. Hauff, K. D. Thiel & B. H. Edwards, photography by R. Lax, A. Jakob & Waitawhile, Wartaweil: Waitawhile.

Theweleit, K. (1998), *Ghosts: Drei leicht inkorrekte Vorträge*, Frankfurt am Main: Stroemfeld/Roter Stern.

Theweleit, K. (2018), 'Diese Körper sind von Angst erfüllt. Interview von Georgios Chatzoudis mit Klaus Theweleit über aktuelle Gewalt in Deutschland', *L.I.S.A. – Wissenschaftsportal der Gerda Henkel Stiftung*, 18 September 2018, online at: https://lisa.gerda-henkel-stiftung.de/maennergewalt_theweleit [accessed 21 December 2018].

Thompson, M. (2017), 'Whiteness and the Ontological Turn in Sound Studies', J. Lavender (ed.), 'Sounding/Thinking', Special Issue of *Parallax* 23 (3): 266–282.

Tilford, K. (2017), 'Generalized Transformations and Technologies of Investigation: Laruelle, Art, and the Scientific Model, Keith Tilford', in R. Gengle & J. Greve (eds), *Superpositions: Laruelle and the Humanities*, Lanham, MD: Rowman & Littlefield, 139–156.

Trier, L. Von (2011), *Melancholia*, Hvidovre Denmark: Zentropa.

Tuin, I. Van der (2018), 'Diffraction', in R. Braidotti & M. Hlavajova, *Posthuman Glossary*, New York: Bloomsbury, 99–101.

Vaihinger, H. (1913), *Die Philosophie des Als Ob. System der theoretischen, praktischen und religiösen Fiktionen der Menschheit auf Grund eines idealistischen Positivismus*, Berlin: Reuther & Reichard.

Valéry, P. (1965), *The Collected Works of Paul Valery – Part V: Idee fixe: A Duologue by the Sea*, translation by D. Paul, Preface by J. Mathews, Introduction by P. Wheelwright, Princeton, NJ: Princeton University Press.

Valéry, P. (2007), *Cahiers/Notebooks 3*, translated by Norma Rinsler, Paul Ryan, Brian Stimpson, based on the French Cahiers edited by Judith Robinson-Valéry, Frankfurt am Main: Peter Lang.

Veen, T. C. V. (2016) 'Robot Love is Queer: Afrofuturism and Alien Love', *Liquid Blackness* 3 (6): 92–106.

Voegelin, S. (2010), *Listening to Noise and Silence: Towards a Philosophy of Sound Art*, New York: Bloomsbury.

Voegelin, S. (2012), 'Ethics of Listening', *Journal of Sonic Studies* 2 (1).

Voegelin, S. (2014), *Sonic Possible Worlds: Hearing the Continuum of Sound*, New York: Bloomsbury.

Wacquant, L. (2015), 'For a Sociology of Flesh and Blood', *Qualitative Sociology* 38: 1–11.

Weelden, D. Van (1999), 'Some Excursions into Sonic Fiction: A two-step with Kodwo Eshun', *Mediamatic Magazine* 9(4) and 10(1), fall, online at: https://www.mediamatic.net/en/page/8733/some-excursions-into-sonic-fiction [accessed 21 December 2018].

Weheliye, A. G. (2005), *Phonographies: Grooves in Sonic Afro-Modernity*, Durham, NC: Duke University Press.

White, H. (1973), *Metahistory: The Historical Imagination in 19th-century Europe*, Baltimore: Johns Hopkins University Press.

INDEX

academic writing/reading,
    fictional/poetic writing in 22–5,
    84–6
acid communism
    Acid Communism (Fisher)
        121–2
    anticipation 109–12
    compulsion 109–12
    defining 121
    embodied theories 116–20
    Eshun and 107
    Fisher and Eshun 107
    ghosts of our times 112–16
    hauntology 109–13, 116, 117,
        120
Acid Communism (Fisher) 121–2
activism, sonic fiction as 144–6
Adams, Douglas 119
affective tonality 23
affect-loaded fictions 114
African Americans, science fiction
    by 8
afrofuturism
    alienation at core of 62–3
    alien continuum 75
    alternate historiographies 80–1
    coinage of term 8
    core experience of 61–2
    defined 72
    diffraction 71–2
    generativity emerging from 62
    and Neoreaction 15
    reaction to cultural phenomena
        15

resistance, example of 64
sonic afromodernity 72
sonic fiction and 3–4
Akomfrah, J. 2, 23
alienation at core of afrofuturism
    62–3
alien nation 75–9
AlterDestiny 75–9, 148
Alter Nation 75–9, 148
anticipation 109–12
anti-essentialism 32
Anzaldúa, Gloria Evangelina 78
apophantic substrate 14
argument, constructivist character
    of 118
artists' theories 28–9
assimilative quality of sonic fiction
    4
audio papers 27
audiopietists 32
audiovisual litany 31–3, 69
aurality, concept of 66–7
    see also black aurality; white
        aurality
autohistoria 78–9, 80–1
Avanessian, A. 59–60, 80

Badiou, Alain 108
Badley, L. 49
Ballard, J. G. 119
Barad, Karen 70, 71
bass cults 138–9
Baudrillard, Jean 13
Black Atlantic 65–6

black aurality 26
  AlterDestiny 75–9
  Alter Nation 75–9
  artefacts, circulation of 66
  aurality, concept of 66–7
  autohistoria 78–9
  Black Atlantic 65–6
  black fugitivity 68–9
  continuous transport as
      characteristic 68
  critique, sonic fiction as 146–8
  decolontologies 80–2
  diffraction of mythscience 70–5
  mistreatment by white aurality
      67
  refusal to be refused 68–9
  resistance, example of 64
  transhistorical given, fallacy
      of 69
black fugitivity 68–9, 126–8
black science, mythscience as 26
black technopoetics 59, 77, 139
Boards of Canada 106
body of the researcher 87–92
breaking the table 123–5, 138
Burial 111
Burroughs, William S. 12, 13

Campt, Tina M. 68
Chude-Sokei, L. 59
Cobussen, M. 28, 29, 30, 31
Collins, Lyn 9
colonialism 64
Colquhoun, Matt 121–2
communism 112–13, 114–16
compulsion 109–12
conceptechnics 28
concept fiction 50–1
concepts, new, appropriation of
    35–6
Confessions on a Dance Floor
    (Madonna), Dath's review
    of 47–9

constructivist character of
    argument 118
Cox, Christoph 30, 32, 132
critique, sonic fiction as 146–8
  see also black aurality
Cuboniks, Laboria 62, 63
Cybernetic Culture Research Unit
    (CCRU) 10–14, 140

Dark Enlightenment (Land) 13
Dath, Dietmar 2, 103
  background as writer 46–9
  Even wrong ideas can be made
      true 58–60
  The Implex. Social Progress:
      History and Idea 50–3
  Karl Marx, writings on 52
  mixillogics of 45–50
  review of Confessions on a
      Dance Floor (Madonna)
      47–9
  translation of More Brilliant
      than the Sun (Eshun) 43–5,
      46–7
  Valéry, origin of implex
      concept and 55–7
Davis, C. 113
De:Bug (journal) 3
decolonization
  of the aural 64
  cultural 77
decolontologies 80–2
Delany, S. 75, 103
Derrida, Jacques 115–16
Dery, Mark 8, 72
dialectics of the implex 50–3, 145
diffraction 148
  afrofuturism 71–2
  autohistoria 78–9
  concept of 70
  in epistemology 70
  of mythscience 70–5
  in research 70–1

dissent, urge for 123–4
Drexciya 72–3
dub virus 28

electronic dance music 139–40
embodied theories 116–20
endeetic surplus 14
Engels, F. 114
epistemologies, sonic
    mixillogics of 27–34
    see also sensory epistemologies
Eshun, Kodwo 2–4, 89, 109, 141
    acid communism 107
    approach to writing about
        sound 83–5
    black futurity 77
    broadening of accepted forms
        of knowledge 38
    concepts, new, appropriation
        of 35–6
    diffraction 72–3
    ethnofuturisms 81
    Fisher, connection with 106–9
    generative epistemology 98
    humanoid experientiality 96–7
    hyperstitions 119
    individual imaginations, role
        of 94
    interpretative communities, list
        of 120
    Mark Fisher Memorial Lecture,
        January 2018 16–17,
        106–7, 118–19, 120
    mutantextures 35, 100
    and Nick Land 12–13
    non-disciplines 133–4
    Otolith Group 16
    polysensory/polyhistoric
        knowledge of music 5–6
    practices, thinking as operating
        through 94–5
    on Rock My Religion
        (Graham) 106

Serres, common/different traits
    with 92–7
technology and 95–6
theoretical tools, test of 90
as thinker of praxis 93
visceral and material,
    arguments rooted in 93
see also More Brilliant than the
    Sun (Eshun)
Essais (Montaigne) 86
essentialism 32
ethnofuturisms 80–2
Even wrong ideas can be made
    true (Dath) 58–60

Fanged Noumena (Land) 13
Fanon, Franz 63
fictional/academic reading/writing
    84–6
Fisher, Mark 15, 16–17, 105
    acid communism 107
    Acid Communism 121–2
    Eshun, connection with 106–9
    Ghosts of My Life 110
    hauntology 105, 111, 116
    hyperstition 118, 119
    lost utopia 109–12, 116, 146–7
    Weird and the Eerie, The
        116–17
    'What is Hauntology?' 109
Five Senses, The (Serres) 87–92,
    144–5
force of liberation, sonic fiction as
    5–10
Fowler, Jarrod 134, 135, 137, 144
fugitivity, black 68–9, 126–8
futurisms 59–60
    see also afrofuturism

Geerts, E. 71
generative epistemology 98–9
Ghosts of My Life (Fisher) 110
ghosts of our times 112–16

Gilroy, Paul 65
goal of sonic fiction 10
Goh, Annie 30
Goodman, Steve 14, 22–7, 39–40,
    41, 147
groups of otoliths 17
Gunkel, H. 77, 81, 102

Halberstam, Jack 127
Hall Nathaniel 19
Hameed, Ayesha 77, 81, 102
Harney, Stefano 68, 76
hauntology 105, 106, 109–13,
    116, 117, 120
Havis, D.N. 64
Hellblau (Meinecke) 73–4
*Heller Als Die Sonne see More
    Brilliant than the Sun*
    (Eshun)
heuristic fictions, sonic fiction as
    143
heuristics of the sonic 149–50
holosonic control 26–7
Holt, Macon 107–8, 110
humanoid experientiality 96–7
hyperstitions 117–19

Idée Fixe (Valéry) 54
idiosyncratic sensibilities,
    mythscience of 36–7
ID Verlag 45–6
imagination, integration into
    research 25
implex
    dialectics of the 50–3, 145
    *Even wrong ideas can be made
        true* (Dath) 58–60
    lost utopia 112
    origin of concept 53–7
*Implex. Social Progress: History
    and Idea, The* (Dath and
    Kurchner) 50–3
individual imagination, integration
    into research 25

Industrial Revolution 57
interpretative communities 17,
    107, 108, 118, 120, 125

Jasen, Paul 99, 119, 138–9, 144
Journal of Sonic Studies 29

Kirchner, Barbara 50–3, 55–6, 57
Koppe, Franz 14, 25
Kösch, Sascha 20
Kraus, C. 2

Land, Nick 10–13, 117, 119
language operation 14
Laruelle, François 129, 130–1,
    133, 139, 140
Last Angel of History, The
    (Akomfrah) 23–4
Lavender, J. 20
Lefebvre, Henri 93
Lehnerer, T. 29
Lem, Stanislaw 119
liberation, sonic fiction as force of
    5–10
Liebezeit, Jaki 149–50
Living in the Moment™ 106
logocentrism
    deviation of through
        mythscience, mixillogic and
        mutantextures 40–1
    sensory epistemologies 97–102
Lorde, Audre 101, 102, 125, 139
lost utopia 109–12, 146–7

Mackay, R. 11–12
Manifesto of the Communist Party
    114–15
Marcuse, Herbert 121
Mark Fisher Memorial Lecture,
    January 2018 16–17,
    106–7, 118–19, 120
Martial Hauntology (Goodman
    and Heys) 27
Marx, Karl

Dath on 52, 58
Manifesto of the Communist Party 114
Meelburg, V. 28, 29, 30, 31
Meinecke, Thomas 73–4, 81
Mestiza Futurity 78
mixadelics 28
multiplying epistemologies 103
mixillogics
  academic status of texts 29
  concept of 21
  Dath's 45–50
  Dath's writing 48–9
  deviation of white science through 40–1
  generating mutantextures 35
  mutantextures and 35
  by practitioners, sonic thinking and 29
  Rock My Religion (Graham) 106
  sensory epistemologies 99–100
  of sonic epistemologies 27–34
  sonic thinking and 147
mixillontology 39
Moalemi, M. 59–60, 80
Montaigne, Michel de 86
More Brilliant than the Sun (Eshun) 89
  approach to writing about sound 83–5
  black futurity 77
  Dath's translation 43–5, 46–7
  diffraction 72–3, 74
  German version, context for 46
  goal of sonic fiction 10
  as long review of theories 20
  mythscience 19
  origin of term *sonic fiction* 2–4
  publisher of *Heller Als Die Sonne* 45–6
  reordering of discourse 5–6

sonic fiction not defined in 5–6, 141
  usage, definition of *sonic fiction* by 6
  see also Eshun, Kodwo
Moten, Fred 63, 64, 68, 76, 126
Mothership Connection 8
multiple epistemologies 98
multiplestomologies 98
Musil, Robert 55
mutantextures
  concept of 21–2
  Dath's writing 48–9
  deviation of white science through 40–1
  Implex. Social Progress: History and Idea, The (Dath and Kurchner) 50
  sensory epistemologies 99–100, 101
  of sonic possible worlds 35–40
  sonic thinking and 147
mythscience 19
  as black science 26
  Dath's writing 48
  deviation of white science through 40–1
  diffraction of 70–5
  holostomic control 26–7
  idiosyncratic sensibilities 36–7
  sensory epistemologies 98–9, 101
  sonic epistemologies and 30
  sonic thinking and 19–20, 27, 147
  of sonic warfare 22–7

Nelson, M. 72
nomad science 99
NON
  breaking the table 123–5
  as marker of resistance 125
  non-musicology 26, 133, 134–8

non-philosophy 129–34
rhythmight 134–8
ultrablack resistance 138–40
non-musicology
    electronic dance music 139–40
    non-philosophy 133
    rhythmight 134–8
ultrablackness 26
non-philosophy 129–34
nontology 76, 80–1

Oliveira, Pedro 77
onotologies, universalist use of
    63–4
O'Sullivan, S. 77, 81, 102
Otolith Group 16
Otoliths 17

Pallat, Nikel 124
Pelleter, Malte 9
personal sensibility, integration
    into research 25
phenomenology 130
Pitts, Andrea J. 79
Plant, Sadie 10
practitioners' theories 28–9
Praxistheorien 28–9

Quest, The (Drexciya) 73

reading, academic/fictional 84–5
refusal to be refused 68–9
resistance
    afrofuturism as example of 64
    black aurality as example of 64
    NON as marker of 125
    sonic, as activism 144
    sonic fiction as example of 64
    ultrablack 138–40
revolutions 57
Reynolds, Simon 11–12, 140
rhythmanalysis 93
rhythmight 134–8
Rock My Religion (Graham) 106

Sagar, A. 109
Schulze, H. 28, 29, 30, 31, 34
science fiction
    by African Americans 8
    non-philosophy 129–34
    in sonic fiction 107–8
science theory fiction 119–20
Scrimshaw, Will 31–2, 33
sense of possibility 55
sensory epistemologies
    body of the researcher 87–92
    Eshun's approach to writing
        about sound 83–5
    everyday experience 97–8
    generative epistemology 98–9
    logocentrism, beyond 97–102
    mixillogics 99–100
    multiple epistemologies 98
    multiplying epistemologies
        102–3
    mutantextures 99–100, 101
    mythscience 98–9, 101
    reading/writing, academic/
        fictional 84–6
    syrrhesis fiction 92–7
    theoretical tools, test of 90
    as transcending framework of
        academia 97
Sensory Mestiza Fiction 78
Serres, Michel 87–92, 144–5
    Eshun, common/different traits
        with 92–7
    humanoid experientiality 96–7
    individual imaginations, role
        of 94
    multiplying epistemologies
        102
    practices, thinking as operating
        through 94–5
    Symposion (Plato) 90–1, 100–1
    technology and 95–6
    as thinker of praxis 93
    viseral and material, arguments
        rooted in 93

Sinker, M. 79
skratchadelia 28
social progress 50–3, 56, 145–6
sociopoetics 60
sonic afromodernity 72
sonic epistemologies 144–5
    mixillogics of 27–34
    see also sensory epistemologies
sonic essentialism 32
sonic fiction
    as activism 144–6
    afrofuturism and 3–4
    appropriate application of
        35–6
    assimilative quality of 4
    as black cultural concept 4
    circulation of 20–1
    concepts, new, appropriation
        of 35–6
    as critique 146–8
    defined 1–2, 142–3
    goal of 10
    as heuristic fictions 143
    as new way thinking about
        music 6
    origin of term 2–4
    resistance, example of 64
    as subjectivity engine 7
    usage, definition by 6
sonic flesh 34
sonic materialism 30–4, 63, 102
sonic possible worlds,
    mutantextures of 35–40
Sonic Possible Worlds (Voegelin)
    37–9, 147
sonic sensibilities 146–7
    see also acid communism
sonic thinking
    critique, sonic fiction as
        146–8
    mythscience, mixillogic and
        mutantextures as core
        concepts 40, 147

mythscience and 19–20, 27
    non-musicology 133
sonic warfare 128–9
    mythscience of 22–7
Sonic Warfare: Sound, affect,
    and the Ecology of Fear
    (Goodman) 22–7
sonocentrism 31–3
sound culture 66, 67
Sprache und Bedürfnis (Language
    and Need) (Koppe) 14
Stadler, G. 4
Sterne, Jonathan 69
subjectivity engine, sonic fiction
    as 7
Sun Ra 75–6, 79, 149, 150
Symposion (Plato) 90–1, 100–1
syrrhesis 73
syrrhesis fiction 92–7, 103, 144–5
Szepanski, Achim 126, 128, 129,
    133, 135, 144

table, breaking the 123–5, 138
technology, Eshun and Serres and
    95–6
technopoetics 59, 60, 77
textual analysis 14
theoretical tools, test of 90
theory-fiction 13–15, 85, 118–19
    Implex. Social Progress:
        History and Idea, The (Dath
        and Kirchner) 50–1
theory writing, multiplying
    options for 119
Theweleit, Klaus 113–14
'Think (About It)' (Collins) 9
Thirst for Annihilation (Land) 13
Thompson, Marie 30, 63, 64
time in fiction 105–6
transhistorical given, fallacy of 69
translated writings 43–5
Tuin, I.V.D. 71, 72
turntabilization 28

ultrablackness
    black fugitivity 126–8
    breaking the table 123–5
    as marker of resistance 125–6
    non-musicology 26
    resistance 138–40
universalist use of onotologies
    63–4
Untrue (Burial) 111
usage, definition of sonic fiction
    by 6
utopia
    acid communism 121–2
    lost 109–12, 146–7

Valéry, Paul 50, 51, 53–7
'Valley of the Shadows' (Pelleter),
    Malte 9

vanilla science 26
Voegelin, Salomé 30, 31, 32, 34,
    36–40, 100, 102, 147, 150

Weheliye, Alexander 72
Weird and the Eerie, The (Fisher)
    116–17
'What is Hauntology?' (Fisher) 109
white aurality 30, 63–4, 67
white science 26
    deviation of through
        mythscience, mixillogic and
        mutantextures 40–1
    repression of diffraction 71
writing, academic/fictional 22–5,
    84–6